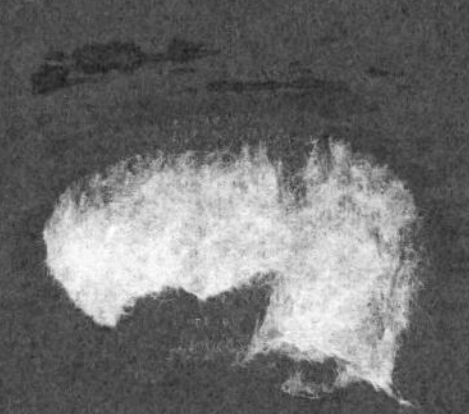

Protozoa

ALBERT WESTPHAL
Professor of Zoology at the University of Hamburg

in collaboration with

HEINZ MÜHLPFORDT
Director of the Protozoology Department of the
Bernhard-Nocht-Instituts für Schiffs-
und Tropenkrankheiten, Hamburg

Translation supervised by
Professor W. H. R. Lumsden
and
Dr G. A. T. Targett
Department of Medical Protozoology
London School of Hygiene and Tropical Medicine

Blackie

Glasgow and London

Blackie & Son Limited
Bishopbriggs
Glasgow G64 2NZ

450/452 Edgware Road
London W2 1EG

593
WES

This work was originally published
in German as *Protozoen* by
Verlag Eugen Ulmer, Stuttgart, 1974

International Standard Book Number

0 216 90216 9

Printed in Great Britain by
Thomson Litho Ltd., East Kilbride, Scotland

Preface
to the English Edition

IN A SCHOOL CONCERNED PRIMARILY WITH TEACHING WHICH IS necessarily restricted to the Protozoa of direct importance to man—because of their economic significance and/or effects on health—one is often sharply conscious of the eclecticism of such teaching and disappointed that students cannot be introduced to the Protozoa as a whole with the immense and fascinating variety of their form, function and mode of life. It was, therefore, with pleasure that we acceded to the British Publishers' request to assist them in the preparation and overseeing of a translation of this book, which seemed to us to fulfil a long-standing need for a concise account in English of the whole Protozoa.

The book, as the authors say, is a general introduction to the Protozoa. Though protozoology, like most disciplines today, is necessarily a subject of specialization and fragmentation, this text is thought likely to be useful to all kinds of specialists and to appeal to a wide audience. It is concise, comprehensive and systematic. It is still, however, highly readable and would provide a useful basis for students of invertebrate zoology, cell biology, ecology and ultra-structure, as well as serving as a work of wider reference.

Even as general a text as this must reflect to a large degree the particular interests of its authors. The emphasis in the present work is on structure, taxonomy and cell organelles in relation to cell function.

In monitoring the English translation of the German text, we have tried to reproduce as faithfully as possible the statements and views of the original authors, and have tried to confine changes solely to those needed for the clarification of meanings or for conformity with English usage.

W. H. R. LUMSDEN and G. A. T. TARGETT
London School of Hygiene and Tropical Medicine.

Preface

BECAUSE OF ITS LIMITED SCOPE THIS BOOK IS NOT INTENDED TO be, and can not be, anything more than an introduction to Protozoology. To this end, the text itself has been kept as short as possible and the numerous illustrations, especially in the taxonomic section, should make meanings sufficiently clear. All the illustrations have been redrawn to achieve a uniform standard of reproduction. I would like to express my grateful thanks to Mrs Susanne Tiepolt for her contribution to this aspect of the book, and also to Mrs Dora Grabosch for some of the illustrations. The illustrations have been chosen firstly to give a general view of the diversity of form, and secondly to act as an aid to microscopic studies by concentrating on the more common species. The legends to the illustrations generally contain information on the occurrence of the species, to facilitate the collection of samples. In the general section, the main aim has been to present the processes involved in the life of the protozoan cell, and their relationship to the individual cell organelles. Short notes on the ecology of the Protozoa, on their uses as indicators of water purity, and on the parasitic Protozoa as causal agents of diseases and infections in man and animals, are intended to convey something of the practical significance of Protozoology.

I should like to thank Mr Roland Ulmer and Dr S. Volk for their willing help in turning my wishes and ideas into printed reality. I am indebted to Professor W. Kloft who urged me to write this book, which I hope will arouse interest in the many-faceted subject of Protozoology.

Albert Westphal

Contents

Introduction

The position of the Protozoa among living organisms

The word "Protozoa" means first animal, primitive animal. The Protozoa are small animal microorganisms. The approximately 25,000 living species of Protozoa described are mostly from 2 μm to less than 1 mm in diameter or, in a few cases among the ciliates and amoebae, up to 3–5 mm. The really primitive foraminiferan *Psammonyx vulcanicus* (figure 39a) can be 6 cm, while the fossil foraminiferan *Nummulites gizehensis* reached 11–12 cm.

The basic element of all animals is the eukaryotic cell, i.e. cytoplasm delimited by a membrane and containing the karyon or nucleus, likewise enclosed in a membrane. The nuclear membrane disintegrates, usually temporarily, only at nuclear division, after the organization of the chromosomes. The Protozoa, as eukaryonts, are therefore clearly distinct from the prokaryotic microorganisms, the bacteria, blue algae and Rickettsia, in which the nuclear substance lies free in the cytoplasm in the form of nucleoplasm or a nucleoid. In further contrast, the viruses, even smaller, are composed solely of sub-cellular informative nuclear substance. They lack cytoplasm, with the enzyme systems which are necessary for metabolism, so that viruses are dependent on the enzymatic functions of the host cells in which they develop.

The concept of the Protozoa is still by no means clearly defined by the distinctions just mentioned. Within the animal kingdom, the Protozoa are generally considered to be unicellular organisms, in contrast to the multicellular Metazoa. Such an antithesis, however, characterizes at best the outward appearance rather than the intrinsic properties. It was consciously by way of introduction that the cell was designated the basic element, and not the fundamental unit, of all animals. Animal cells are not homogeneous; they can differ from one another both in morphology and physiology. The various morphological differentiations in the structure of the cell are designated **cell organelles**. The structure of the nucleus can also show considerable differentiation. Nuclei with a single chromosome complement, with one genome, are called **haploid**; those with a double complement are called **diploid**; and those possessing several genomes, **polyploid**. One cell can contain one or numerous nuclei. If several nuclei are present, these can be similar or dissimilar, e.g. diploid and polyploid. All these possibilities occur in the Protozoa. But there is also a conspicuous abundance of cytoplasmic differentiation. The Metazoa possess numerous organs for the different functions of the body, which are formed from specifically organized tissues and are interdependent. The Protozoa must fulfil all functions without the ability to form tissues. Protozoa are animals without tissues. Therefore they develop to only a small overall size. However, they can—particularly in some flagellates and ciliates—show a degree of differentiation of cell structure never achieved in the cells of the Metazoa.

Although the Protozoa can be distinguished relatively clearly from the Metazoa, at least in their developmental characteristics (although intermediate forms may also arise here), the distinction from plant microorganisms is more difficult. The algae too are unicellular eukaryonts. The differentiation between animals and plants depends on the nutritional physiology of the cell. In specific pigment carriers, the plastids (chloroplasts), the true plant algae possess assimilatory ability which is activated under the influence of light energy. They are phototrophic and in that they are not dependent on substances from other organisms for their nutrition but need only inorganic compounds for their metabolism) also autotrophic. The truly animal Protozoa, on the other hand, are heterotrophic. They are dependent on organic substances or the products of their breakdown. Whilst the majority of Protozoa are heterotrophically maintained, there are amongst the flagellate Protozoa, the Flagellata or Mastigophora, numerous species which are adapted to autotrophic and heterotrophic life. They are generally termed **mixotrophic** and are thus both plants and animals; Haeckel introduced the term **protists** for such organisms. For the protists, the classification as animal or plant has not yet been made. The precariousness of nutritional physiology as a means of distinguishing between animals and plants is shown also by the fact that the eukaryotic fungi as well as the prokaryotic bacteria are heterotrophic. Furthermore, many flagellates can be either autotrophic or heterotrophic, according to environmental factors.

Principles in the classification of the Protozoa

The classical taxonomy of the Protozoa is based on the nature of their locomotory organelles and on the formation of encapsulated resistant stages. The flagellates and ciliates are characterized by the possession of filiform oscillating locomotory organelles. These are relatively long in the flagellates, mostly longer than the rest of the cell, and are called **whips** or **flagella** (figure 1a). The locomotory organelles of the ciliates are generally shorter than the body of the cell itself, and are distinguished from flagella as **cilia** or hair-like threads (figure 1c).

The Rhizopoda show peripheral areas of cytoplasmic streaming which protrude for a time from the protoplasmic body and can then be reabsorbed. Such locomotory organelles are called **pseudopodia** or false feet (figure 1e). All these locomotory organelles function only in the vegetative stages of the Protozoa, the **trophozoites**. In contrast to the vegetative motile stage, many Protozoa can surround themselves with an outer capsule. This immotile form is called the **cyst** or perennial state (figure 1b, d, f). Through the formation of cysts, the Protozoa become largely resistant to adverse environmental conditions. Vegetative cells are released once more from the cyst by the discarding of the empty cyst membrane.

The Sporozoa are characterized, not by their locomotory organelles, but by their ability to produce spores. All Sporozoa are parasites, often with complex developmental cycles. Spores, surrounded by a strong coat, may form either within cysts, whose formation begins with the fusion of sexually differentiated stages, or within multicellular mother cells (figure 1 g, h). Constrictions in these mother cells may also lead to spore

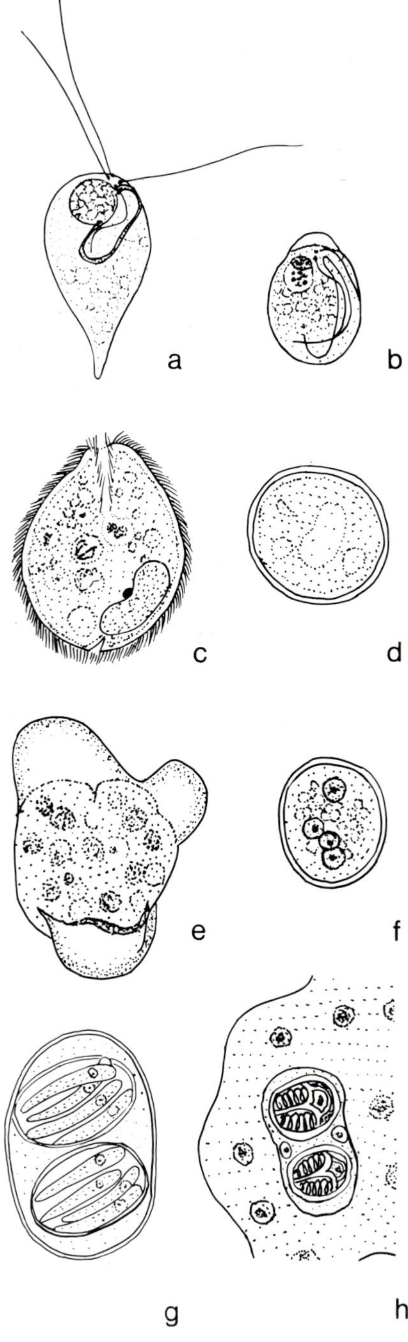

formation. Germ nuclei, the sporozoites, develop in the spores and then achieve further development in a new host, thereby beginning a new infection cycle.

These principles of the systematic classification of the Protozoa are still valid even today, although they leave the position of many species unclear. Supplementary criteria for the systematic classification of the Protozoa—such as reference to the nuclear dimorphism of the ciliates, which form a micronucleus and macronucleus, or even structural analyses by electron microscopy—have given only limited help in taxonomy. It is not surprising that questions still remain open since, on account of the perishable nature of the Protozoan structure, apart from some skeletal and shell structures, we have access only to present-day material; palaeontological sources illuminating the history of the phylum, the phylogeny, are practically nonexistent. For this reason, the ontogeny, i.e. the course of development of the extant species, has a particular significance in determining systematic relationships. This applies particularly to the scheme of development of the Sporozoa. For the classification of the Flagellata, the presence or absence of plastids constitutes an important fundamental criterion; other criteria are the complement of flagella and the character of the nuclear apparatus. The Rhizopoda are distinguished on the basis of their pseudopodial formation, as well as on their protective and supportive organelles. Characteristics with taxonomic significance in the Ciliata, besides the complement of cilia, are the structure and position of the cytostome, the cell-mouth,

---

Figure 1. **The Classes of the Protozoa**
a–b **Flagellata, Mastigophora**
   a trophozoite of *Chilomastix mesnili,* **b** cyst
c–d **Ciliata, Ciliophora**
   c trophozoite of *Balantidium coli* (30–150 µm), **d** cyst
e–f **Rhizopoda**
   e trophozoite of *Entamoeba histolytica* (tissue form with phagocytozed erythrocytes)
   f mature cyst
g–h **Sporozoa**
   g oocyst of *Isospora bigemina,* with 2 sporocysts, each with 4 sporozoites
   h spore enclosure (sporont, sporoblast) with 2 spores, each with an amoebula and two spore capsules, in a section of the multinucleate plasmodium of *Myxobolus pfeifferi*

and also the special nature of the nuclear apparatus and the particular type of sexuality. The nuclear dimorphism of the typical ciliates has led to the placing of the Ciliata in a special division, Cytoidea or Heterokaryota, separate from the division Cytomorpha or Plasmodroma in which the Flagellata, Rhizopoda and Sporozoa were combined. To stress this division is now no longer justified, since it has been demonstrated that a nuclear dimorphism can occur also in the Foraminifera belonging to the Rhizopoda.

As a result of new knowledge gained from time to time, and of the different values placed by different individuals on basic taxonomical characteristics, new classifications are for ever being proposed within the individual classes of the Protozoa. In doing this, we must also take into consideration whether the systematic arrangement is to serve as a general introduction to proto-zoology or as a tool for specialists. The latter are referred particularly to the classification which appears in the communication from the Society of Protozoologists;* the scope of this book is such that an extensive key cannot be included. The majority of proposed systems subdivide the phylum Protozoa into four classes, as mentioned above, although there is also a proposal to classify the Protozoa as a sub-kingdom divided into four phyla.

* see bibliography, page 309.

Animal Kingdom
Phylum 1 : Protozoa

CLASS I: FLAGELLATA (COHN 1853: WHIP ANIMALS)
OR MASTIGOPHORA (DIESING 1865: WHIP BEARERS)

SUBCLASS 1: PHYTOMASTIGOPHORA (Calkins 1909)

The Phytomastigophora or Phytoflagellata are flagellates with
pigment-bearing bodies which are generally called **chromatophores** or **plastids**. The Phytoflagellata are autotrophic or
mixotrophic. If the mixotrophic flagellates can maintain themselves in the dark purely heterotrophically, they are also
designated **amphitrophic**. Some heterotrophic flagellates are
also assigned to the Phytomastigophora if their relationship with
pigmented flagellates is morphologically apparent. By no means
all pigmented Phytoflagellata are green in colour. With other
kinds of pigmentation the plastids are called **chromoplasts**, to
distinguish them from the chloroplasts which are green.

Order 1: Chrysomonadina (Stein 1878), figures 2 and 3

The name Chrysomonadina means gold-celled, and refers to the
predominantly yellowish or brownish chromoplasts. A red eyespot (stigma) can occur at the base of the flagellum. Fresh-
water species often possess a contractile vacuole. Leucosin,
which is very shiny in fresh preparations and which is probably
a carbohydrate, is deposited as a characteristic metabolic
product, as are oils and fats but never starch. Colony formation
is frequent. Some species can also produce pseudopodia or can

completely transform to the amoeboid condition. Chryso-monadina are predominantly fresh-water Protozoa. The following can be distinguished:

Chromulinidae, with only one terminal flagellum (figure 2 a–e, figure 3 a–b).

Isochrysididae, with two equal flagella (figure 2 f–h, figure 3 c).

Ochromonadidae, with two unequal flagella, one main and one subsidiary (figure 2 i–m, figure 3 d–f).

Coccolithophoridae, with a shell of assembled coccoliths (small plates of calcium carbonate), mainly marine (figure 2 n, figure 3 g–h).

Silicoflagellidae, marine, with a skeleton of hollow silica spicules (figure 2 o–p).

Rhizochrysidae, which are completely different in that they lack flagella and only the amoeboid state occurs (figure 2 q); if chromatophores had been absent, they would have been placed in the Rhizopoda.

Order 2: Heterochloridina (Pascher 1912), figure 4

Relatively few species are included in the Heterochloridina which, as their name suggests (unequal light-green), possess yellowish-green chromatophores. Just as in the Chrysomonadina, the cyst coat arises first peripherally within the cell body. It comprises two halves of a silicate shell. The typical species have a main and a subsidiary flagellum. The Heterochloridina also include species which occur mostly, or indeed solely, in the amoeboid state. They occur in fresh water or sea water. Leucosin and oil are synthesized as reserve substances.

Order 3: Cryptomonadina (Stein 1878), figure 6

In the typical Cryptomonadina ("hidden-celled" forms) the cell mouth lies at the end of a funnel-shaped gullet which arises as a continuation of an anterior ventral groove. The Crypto-monadina are asymmetrical, so that a ventral and a dorsal aspect can be distinguished. In the gullet wall lie ejectisomes (see figure 137 e), which are ejected by chemical stimuli. From the base of the gullet arise two somewhat unequal flagella, which can pass as a rhizoplast in the cell cytoplasm as far as the nucleus. The chromatophores are of various colours (yellowish, reddish, brown, bluish or green). There are diverse products of assimilation, such as pure starch or amyloid carbohydrates, as well as fat and oil. The nutrition is autotrophic and mixotrophic and, in some species which have no chromatophores (see figure 6b), also heterotrophic. Some species possess a stigma. The contractile vacuole is situated towards the front, at times also towards the back in the cell. Amoeboid forms occur rarely. Fresh-water and salt-water species.

Order 4: Dinoflagellata (Bütschli 1885), figures 5, 7, 9

The word Dinoflagellata means "whip-curl animalcules". Of the two flagella which arise side by side, one curls transversely round the cell like a girdle, whilst the other is free behind as a longitudinal or trailing flagellum. In typical forms, the flagella lie in grooves in the cell. Dinoflagellata are abundant in the sea, but also occur in fresh water. The forms living near the surface usually have many small brown chromatophores, and are autotrophic or mixotrophic. Many have a stigma. The deep-sea parasitic species are colourless and heterotrophic. Dinoflagellata can be coated in armour or naked. The strong coating consists of cellulose and organic deposits penetrated by pores for the fine pseudopodia. In the **Prorocentridae**, the shell comprises two valves united at a seam. The flagella arise atypically, from the anterior pole (figure 5c). In the **Dinophysidae**, winged ridges arise laterally in the region of the sagittal seam of both halves of the shell; the longitudinal flagellum passes between them. At the anterior end, two annular wings form the groove for the transverse flagellum (figure 5a–b). The **Peridinidae** possess shells which are divided up into individual armour plates (figure 5d–e, figure 7a–b). The **Gymnodinidae** have no strong shell formation. The flagella lie in transverse and longitudinal grooves on the cell (figure 5f–o, figure 7c–e). The Gymnodinidae show a wide range of morphological variation. A stigma structure with a lens in front can occur in deep-water forms (figures 5f, 139a). In some species, a multiplication of all cell organelles leads to the formation of colonial individuals (figure 5k).

Parasitic species possess the outward appearance of the Dinoflagellata only when they are immature swarmers (figure 5l–o, figure 7e–g), but a change in form such as this is also possible among free-living species (figure 9a–b). The Dinoflagellata synthesize fats and to some extent also starch as reserve substances.

Order 5: Euglenoidina (Blochmann 1895), figures 8, 10

The Euglenoidina ("beautiful-eyed" forms) derive their name from the red, yellowish or brownish eye-spot of the green pigmented species. The form of the Euglenoidina is usually elongate, often with a screw-like twist, and more or less strengthened by a spirally textured surface (pellicle). There is a sac-like depression at the anterior end, from the bottom of which arise one or two flagella. This small flagellum-sac serves also as a cytopharynx, i.e. a cell gullet, since most forms are mixotrophic and hence also take up solid food. The species without chromatophores are true heterotrophs. The Euglenoidina are predominantly freshwater Protozoa. The starch-like substance paramylum is synthesized as a food reserve. The **Euglenidae** comprise the pigmented species and other forms similar to the pigmented types but without chromatophores (figure 8a–m, figure 10a–c). The **Peranemidae** are colourless Euglenoidina which have no parallel forms among the pigmented species (figure 8n–r).

Order 6: Chloromonadina (Klebs 1892), figure 11

The Chloromonadina ("blue-green celled") constitute a small group of Flagellata with bright green, small, discoid chromatophores which turn blue-green with the addition of acid. Oils are laid down as metabolic products, but never starch or starch-like carbohydrates. The Chloromonadina resemble the Euglenoidina in morphology. They are found mainly in freshwater mud.

Order 7: Phytomonadina (Blochmann 1895), figures 12–15

The Phytomonadina ("plant-celled" forms) are also largely mixotrophic. The usually cup-shaped plastids are grass-green. Pure starch is synthesized as a food reserve, as in the higher plants. Colourless species are heterotrophic and saprozoic. Amoeboid forms occur only rarely. The Phytomonadina are generally small, but in some species regularly form larger colonies. The **Polyblepharididae** have no cellulose membrane. Most of them have four or more flagella (figure 12 a–b, figure 13 a–b). They are predominantly marine. The **Chlamydomonadidae** usually have two flagella and a closely applied or spreading cellulose membrane. They live chiefly in fresh water (figure 12 c–f, figure 13 c–f). The individual cells of the **Volvocidae** conform to the Chlamydomonadina, with green chloroplasts and a stigma. Cell multiplication leads to colonial groups in which morphological and physiological cell differentiation occurs in some species (figure 14 a–e, figure 15 a–e). The Volvocidae occur only in fresh water.

SUBCLASS 2: ZOOMASTIGOPHORA (Calkins 1909)

The Zoomastigophora are flagellates which never have chromatophores; a relationship between them and the pigmented species can no longer be discerned. From the point of view of nutritional physiology, they are true animal flagellates.

Order 1: Protomonadina (Blochmann 1895), figures 16–18

The Protomonadina, the "first-cells", include a very varied range of heterotrophic flagellates. Close relationships exist within the individual families, with the exception of the **Eumonadidae** which contains extremely varied species (e.g. figure 17 a). The **Cercomonadidae** are flagellates with one or more flagella and with amoeboid movement. If more than one flagellum are present, one serves as a trailing flagellum. Forms with no flagella can be completely amoeboid. The Cercomonadidae live in marshes and stagnant water, and some are intestinal parasites (figure 18 a–b). The **Craspedomonadidae**, also called Choanoflagellata, are uniflagellate freshwater flagellates which have a plasmatic collar structure (collare) at their anterior end. The flagellum is free and originates at the centre of the funnel-shaped collar. An outer shell formation can enclose the flagellates in a strong matrix, and a stalk-like growth of the shell can even lead to development of branched colonies (figure 16 a–c). The **Bicoecidae** are biflagellate, the trailing flagellum serving to anchor them in the matrix (figure 16 d). The **Trypanosomatidae** are uniflagellate parasitic flagellates which possess, in the region of the base of the flagellum, a kinetoplast which can show intense pigmentation with certain stains; this is an autonomous organelle which is produced only by replication. The Trypanosomatidae undergo changes in structure (polymorphism). In the promastigote form the kinetoplast lies towards the anterior end of the cell (figure 17 b), in the epimastigote form near the nucleus (figure 17 c), and in the trypomastigote form in the posterior part of the cell (figure 17 d, figure 18 c). The amastigote stage (figure 18 d) has no flagellum; only the nucleus and kinetoplast remains in a small oval cell. In the epimastigote and trypomastigote forms, the flagellum, before it becomes free anteriorly, is joined to the cell body and thereby forms an undulating membrane. The **Bodonidae** are flagellates with a main and a trailing flagellum. Near the base of the flagella the Bodonidae possess a round or oblong kinetoplast. In parasitic species the trailing flagellum can form an undulating membrane (figure 17 e–g, figure 18 e). The **Retortamonadidae** are parasitic flagellates with an anteriorly placed lateral cytostome structure through which a trailing flagellum runs. A further one or three flagella extend anteriorly (figure 17 h, figure 18 f). The **Tetramitidae** are flagellates whose mitotic nuclear division is characterized by peculiar behaviour of the inner body of the nucleus. The Tetramitidae tend to adopt an amoeboid phase; some have become true amoebae whose relationship with the Tetramitidae is only manifest in the peculiar mitosis in which the inner nuclear body (the nucleolus) also participates in the division. The Tetramitidae inhabit decaying organic matter (figure 16 e–g, figure 17 i–l). The **Distomatidae** are free-living and parasitic flagellates which can be called **double individuals**.

All organelles are doubled, so that the distomatids possess two nuclei and two groups of flagella; the free-living species also have two mouth-structures; the parasitic forms generally have no cytosomes and do not take up solid food. Most have eight flagella, though some have only six (figure 18 g–h).

Order 2: Polymastigina (Blochmann 1895), figures 19–21

The Polymastigina ("many-flagellate" forms) are multiflagellate parasitic flagellates which are characterized by the possession of a parabasal body which can be demonstrated by osmium fixation, and also by the occurrence of a peculiar type of nuclear division in which the spindle arises between the centrosomes outside the nucleus (figure 146 g). Only the Pyrsonymphidae have no extra-nuclear central spindle and no parabasal body. The Poly-mastigina are very widely distributed intestinal parasites of arthropods and vertebrates. In xylophagous termites and cock-roaches, species occur whose cell organization attains the highest degree of differentiation known in animal cells (figure 21). The **Trichomonadidae** have 4–6 flagella, of which one serves as a trailing flagellum and—as in *Trypanosoma*—one can form an undulating membrane. In this region the cell is strengthened by a basal fibril (costa) which begins anteriorly at the base of the flagellum. An elastic hardened axial rod (axostyle) traverses the body of the cell antero-posteriorly (figure 19 a–e, figure 20 a–b). The **Calonymphidae** represent colonial individuals which are derived from the Trichomonadidae by a multiplication of cell organelles including the nucleus (figure 20 c–d). The structure of the **Pyrsonymphidae**, too, corresponds to the basic type of the Trichomonadidae, although here all flagella are similar. Further, the central spindle of the centrosomes is an intra-nuclear and not an extranuclear formation during nuclear division, and the parabasal body (which would otherwise have been situated at the source of the flagella) is absent. The species are uninucleate or, where there has been multiplication of the organelles, multinucleate (figure 20 e). In the **Hyper-mastigidae** the flagella and axostyle have multiplied, yet all Hypermastigidae remain uninucleate (figure 20 f, figure 21). In the most highly developed forms, differentiation at the anterior end has produced a head-like structure (operculum) resembling a skull-cap with its dome filled with liquid. This has a flexible connection, the rostral tube, with the flagellate rostrum (figure 21).

The Biology of the Flagellates

Pascher placed the autotrophic Flagellata at the lowest point in the system of evolution of the Protozoa, since their existence is not dependent on other organic substances. However, after Miller (1953) in particular had demonstrated experimentally that, under conditions which existed on our earth some 3 thousand million years ago, the aminoacids important for the heterotrophic organisms could also arise abiotically, Pascher's reasoning was undermined by conclusive evidence. In spite of this, it appears useful, with regard to protozoology, to place the pigmented Flagellata at the beginning of the system, since animal and plant development can be seen to derive from these protista. Many of these flagellates occur in the vegetative phase not only as flagellate organisms, but also as the aflagellate **palmella stage,** which is usually embedded in a gelatinous mass. Unlike cysts, the amastigote palmellae possess a complete metabolic system and the capacity for reproduction (figure 22 a–d). Consequently the palmella stages of the pigmented flagellates behave just like the corresponding amastigote stages of the Trypanosomatidae which lack chromatophores (figure 18 d, figure 23 b). The palmella formation can be seen as the point of departure for the various algae. But other groups of Protozoa, too, can be derived from the phytoflagellates. We have repeatedly seen already transition to the amoebic form (figure 2 b, c, figure 4 e, figure 16 f, g, figure 17 l, figure 18 a). The Heliozoa, which also belong to the Rhizopoda, can be derived from the Chrysomonadina. The dinoflagellate genera **Gymnaster** and **Gyrodinium** show the formation of silica skeletons and central capsules which represent features of the Radiolaria. The Sporozoa, like the Foraminifera, are reminiscent of the Dinoflagellata in their biology, with asexual and sexual reproduction. The holotrichous ciliate *Opalina* is these days often classified with the Polymastigina. Consequently, there are sufficient grounds for placing the Flagellata at the beginning of this survey.

It has already been mentioned that, because of the fragility of the protozoan body, palaeontological findings which would give information about the phylogeny of the Protozoa are generally not to be found. A survey of present-day forms often allows recognition of analogies between the individual groups of Protozoa without indicating the course of phylogeny. The systematic classification of the Protozoa thus remains largely hypothetical. Even the distinction between the Phytomastigophora and Zoomastigophora proposed by Calkins does not imply a clear-cut delimitation. Species lacking chromatophores are often assigned to the mainly mixotrophic Phytomastigophora if other characteristics justify this. Species with plastids can change their appearance. In addition to chlorophyll, other pigments, such as carotin, xanthophyll, phycopyrrin, haematochrome, determine the colour of the flagellates, and some species of Protozoa can vary in colour in response to different physiological conditions. For example, *Haematoccus pluvialis* (figure 13 f) is normally green. When nitrogen and phosphorus deficiencies occur in a volume of water which has collected after heavy falls of rain, red haematochrome predominates (figure 22 e), as often occurs in older cultures, thereby leading to the phenomenon of "raining blood". During the rainy season, *Euglena sanguinea* gives some alpine snows a red colour. Nitrogen deficiency turns *Euglena gracilis* (figure 8 b) orange. The red pigmentation of *Chlamydomonas nivalis* can cause "bloody snow" in some mountain snowfields. Many flagellates, especially the euglenoids such as *Euglena gracilis*, become colourless when maintained in the dark in solutions with an adequate supply of organic substances. Their chromatophores become leucoplasts. They then resemble the permanently colourless genus *Astasia* (figure 8 d–f). Closer investigation shows, as illustrated by the example of *Ochromonas danica* in the Chrysomonadina (figure 22 f, g), that the plastids, which in this case lie close to the nucleus, not only lose their pigment but are also reduced to small rounded proplastids. After a return to a normal environment, they can give rise again to pigmented plastids of the original size.

Reproduction and Sexuality in the Flagellates

Normally, reproduction in the Flagellata proceeds by means of **binary fission** which takes the form of longitudinal division in the flagellate stages (figure 23a). Nuclear division precedes division of the cytoplasm. The flagellum does not divide, the daughter animal producing a new flagellum after the formation of a new basal body. In multiflagellate species, the flagella may first be divided between the two daughter cells, the missing flagella being produced later (figure 23c, d). If the species concerned are not only multiflagellate but also multinucleate, binary fission can proceed as **plasmotomy**, in which the nuclei and flagella present are divided between the daughter animals, and only subsequently are brought up to the full complement (figure 23e). In many species, multiplication of the nucleus and the flagella can also occur by means of multiple divisions before the division of the cytoplasm begins. This type of reproduction is known as **multiple division** or **schizogony** (figure 23g). Binary fission (figure 22a, 23b) or multiple division (figure 23h) can also occur in the amastigote phase. In species with a hard covering formed from jointed plates, the direction of the division follows the line between the plates (figures 23i, 25l). In the binary fission of some flagellates a twisting movement (figure 23f) accompanying the separation of the daughter cells can also give rise secondarily to what appears to be a transverse division. The longitudinal division appearance is similarly masked when the cytoplasm is embedded in a gelatinous matrix, as is the case in the dinoflagellate *Noctiluca miliaris* (figure 23k). Some flagellates which have lost their flagella produce division cysts within which the process of division takes place (figure 23l). In the course of this, the emergent daughter cells can rotate as they separate within the division cyst; here again the impression is given of a transverse division (figure 23m, n). In the palmella stage these division cysts are embedded in a common gelatinous matrix (figure 22d).

Special processes of division occur in parasitic Dinoflagellata. In *Haplozoon clymenellae,* an intestinal parasite of polychaetes (figure 51–o), only the young swarmers have the appearance of dinoflagellates. They attach themselves in the gut of their host by means of a spike, become aflagellate, and grow to a utricular shape, the trophocyte, which becomes multicellular through repeated **budding** and assumes a vermicular outline. The terminal cells, which are called **sporocytes**, become detached and revert to flagellate swarmers. In the dinoflagellate *Blastodinium spinulosum,* an intestinal parasite in copepods (figure 7 e–g), the flagellate swarmer likewise loses its flagellate form and becomes an elongated aflagellate trophocyte. New trophocytes and morphologically similar gonocytes arise by repeated divisions within this trophocyte; the sporocytes are produced from the gonocytes by further divisions. This type of reproduction is known as **palisporogenesis**. Figure 7 g shows a stage with 2 trophocytes and 64 sporocytes. The latter are freed as flagellate swarmers by rupture of the mother cell, and lead to the infection of new host animals.

Apart from the agamous processes of division described so far, many flagellates also have the capacity to undergo a division in which sexually differentiated **gametes** are formed. Such gametes can fuse with each other in pairs; during this process both nuclei also become united (**karyogamy** or **synkaryon formation**). The cell produced by the **fusion** of the gametes is the **zygote**. Figure 24a shows the onset of fusion of two gametes of *Scytomonas pusilla* (Euglenoidina). Since both gametes are of equal size, this is a case of **isogamy** and, inasmuch as the gametes are the same size as the normal vegetative flagellates, this is at the same time **hologamy**. The zygote (figure 24 c) in this case soon becomes enclosed in a cyst and thereby becomes a **cystozygote**. In the example shown in figure 24 d, numerous smaller gametes are developed (figure 24 e) within the flagellate *Chlamydomonas steini* (Phytomonadina) after the encystment of the vegetative stage. These are freed as biflagellate gametes (figure 24 f) and they fuse in pairs (figure 24 g). Since both of the fusing gametes are of equal size, this too is isogamy, but not hologamy, on account of the difference between their size and that of the normal trophozoite. Figure 24h shows the product of the fusion of the gametes, namely the zygote. If the fusing gametes differ from each other in size, the **micro-gametes** are called male and the **macrogametes** female. Figure 24 i shows an **anisogamous fertilization** of this sort between two gametes in *Chlamydomonas brauni*. Comparison with the example of *Chlamydomonas steini* (figure 24 g) shows that both isogamy and anisogamy can occur, even within one genus. If the differences between the microgametes and macro-gametes are even more extreme, the anisogamy becomes **oogamy**, an example of which is illustrated here by *Eudorina elegans* of the Phytomonadina (figure 24k). The slender motile microgametes swarm round the large nutrient-rich immotile macrogametes of the female colony (here represented by a sector) and fuse with them just as a spermatozoon fertilizes an egg.

Sexuality is by no means a basic phenomenon in flagellate development. In many species only **agamogony**, asexual reproduction, occurs. Occasionally only a few species within a particular group show sexuality. Even in species with a sexual phase of development, **gamogony**, this path is only occasionally followed (figure 24 l). In addition, gamogony sometimes occurs only at particular times. In some Polymastigina from the xylophagous cockroach *Cryptocercus punctulatus* and from xylophagous termites, gamete formation of the parasite is induced by the moulting hormone of the host. Figure 24 m shows the formation of two at first only slightly dissimilar gametes in an encysted flagellate of the genus **Trichonympha** in the gut of the cockroach *Cryptocercus punctulatus*. After the gametes have been freed the female gamete develops posteriorly a receptive protuberance, through which the microgamete can penetrate into the macrogamete (figure 24 n, o). The microgamete disintegrates so that only its nucleus remains and this accomplishes karyogamy with the nucleus of the macrogamete.

The development of some parasitic dinoflagellates is worthy of particular attention with respect to further phylogenetic relationships. The special capacity for metamorphosis of the dinoflagellates is already apparent from the two examples already cited, *Haplozoon clymenellae* (figure 51–o) and *Blastodinium spinulosum* (figure 7 e–g). Whilst the swarmers of *Haplozoon* and *Blastodinium* assist in agamous reproduction, it has been shown that swarmers of the genus *Coccidinium* also have the capacity for fusion. The formation of the swarmers is here initiated by schizogony. In *Coccidinium duboscqui*, a parasite of the dinoflagellate *Peridinium balticum*, the parasite is uninucleate at first (figure 25 a). Numerous nuclei arise through multiple division and become arranged around the periphery of the cytoplasm (figure 25 b, c) which is now surrounded by a covering layer; they lead to the formation of microswarmers (d). The swarmers are released through extended utricular structures, the **sporoducts**. Macroswarmers (f–h) are produced by an analogous process, schizogony, again from uninucleate trophozoites which are distinguishable by their high nutrient content (e). Since some species of the genus *Coccidinium* form up to 4 morphologically distinct types of swarmer, it can be assumed that these swarmers, which are also designated **dinospores**, are of differing significance. They could be agametes, like the swarmers of *Haplozoon* and *Blastodinium*, or also represent gametes. That the latter possibility exists has been established for the swarmers of *Coccidinium mesnili*, a parasite of the dinoflagellate *Crypto-peridinium foliaceum*, through the demonstration of fusion and zygote formation (figure 25 i). The further development of the swarmers of this solely parasitic species of *Coccidinium* has not as yet been worked out.

In *Noctiluca miliaris* (figure 9b), a species of the Dino-flagellata which lives free in oceanic plankton, production of swarmers also occurs (figure 9a) in addition to binary fission (figure 23 k). These swarmers arise on a skull-cap-shaped sector of the cell surface (figure 25 j) after multiple division of the nucleus. When the capacity for fusion and zygote formation is demonstrated for these swarmers too, thereby indicating their gametic nature, many authors suggest that agamous swarmers also may arise in the course of schizogony. As with the *Coccidinium* species, the further development of the swarmers and zygotes is also as yet unknown. The detailed studies of cultures of *Glenodinium lubiniensiforme*, a fresh-water dino-flagellate species belonging to the Peridinidae, are therefore all the more valuable. In this species, an agamous binary fission, which proceeds equatorially on account of the protective plating, normally occurs (figure 25k–m); the daughter cells complete the missing half of cytoplasm and outer cuirass as well as the complement of flagella. After 6–7 divisions, during which the flagellates become larger and larger (figure 25 n), a reduction in cell size follows with the formation of two agamous swarmers (o, p). The production of the new armour plating of the swarmers begins only after their release. The development of the swarmers which turn out to be gametes (q–t) occurs only if appropriate populations come together. In this case, 4 gametes arise by means of two divisions; they are released as flagellated swarmers by rupture of the armour plating. Only swarmers from different mother cells fuse with one another (t) to produce the zygote (u). After a few days the zygote membrane is ruptured by secretion of a gluten (v). Two agamous divisions occur within the re-maining glutinous matrix (w, x), and two of the 4 swarmers formed degenerate (y), so that only two are released. These develop their armour plating (z) within a few hours and grow to become normal vegetative stages. This example not only shows how different types of swarmers can arise, but at the same time demonstrates the diversity of developmental paths in the dinoflagellates.

Figure 2. **Chrysomonadina** (Magnification 550 ×)
a–e **Chromulinidae**, uniflagellate:
  a *Chromulina freiburgensis,* with large chromatophore, with contractile vacuole at the front
  b *Chromulina pascheri,* also amoeboid stage with chromatophore
  c *Chrysamoeba radians,* flagellate stage; amoeboid form with pseudopodia; flagellate amoeba stage = mastigamoeba
  d *Ciliophrys marina,* colourless, marine. Heliozoan stage with fine pseudopodia (axopodia) and flagellate form
  e *Chrysopyxis cyathus,* attached, with receptacle
f–h **Isochrysididae**, with 2 equal flagella:
  f *Derepyxis amphora,* with receptacle; attached
  g *Synura uvella,* flagellate with silica scales; colony formation
  h *Rhipidodendron splendidum,* colony with escaping flagellates in some places and the residual empty gelatinous tubes
i–m **Ochromonadidae**, with main and subsidiary flagellum:
  i *Ochromonas fragilis,* lower form with pseudopodium
  k *Stylopyxis mucicola,* solitary, with casing
  l *Dinobryon sertularia,* free-swimming colony in a cellulose casing
  m *Uroglenopsis europaea,* free-swimming colony in a gelatinous sphere (section only)
n **Coccolithophoridae**:
  n *Hymenomonas roseola,* surrounded by annular coccoliths; fresh-water species
o–p **Silicoflagellidae:**
  o *Distephanus speculum,* latticed silica skeleton forming a hemisphere
  p *Dictyocha speculum,* nucleus and chromatophores shown within the silica skeleton
q **Rhizochrysidae**, amoeboid, never flagellate:
  q *Rhizochrysis scherffeli,* amoeba stage with chromatophore

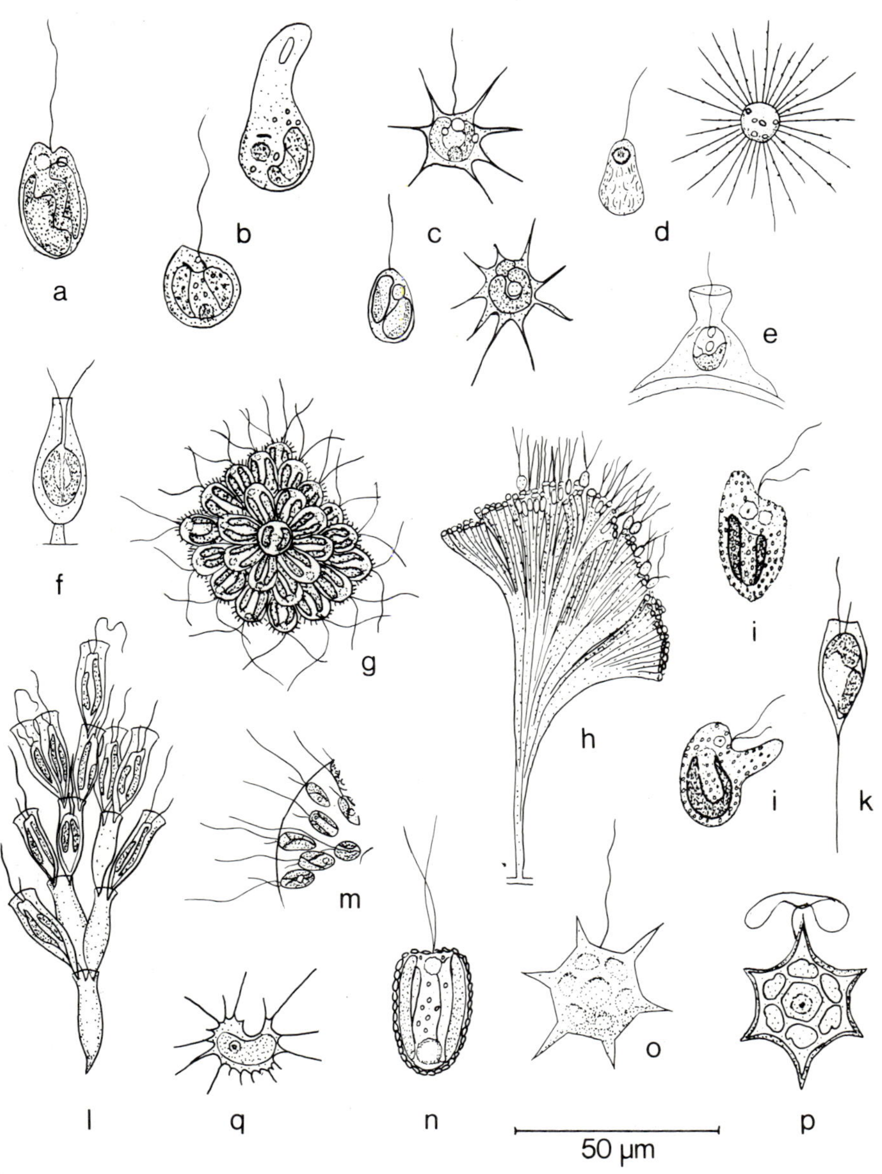

50 μm

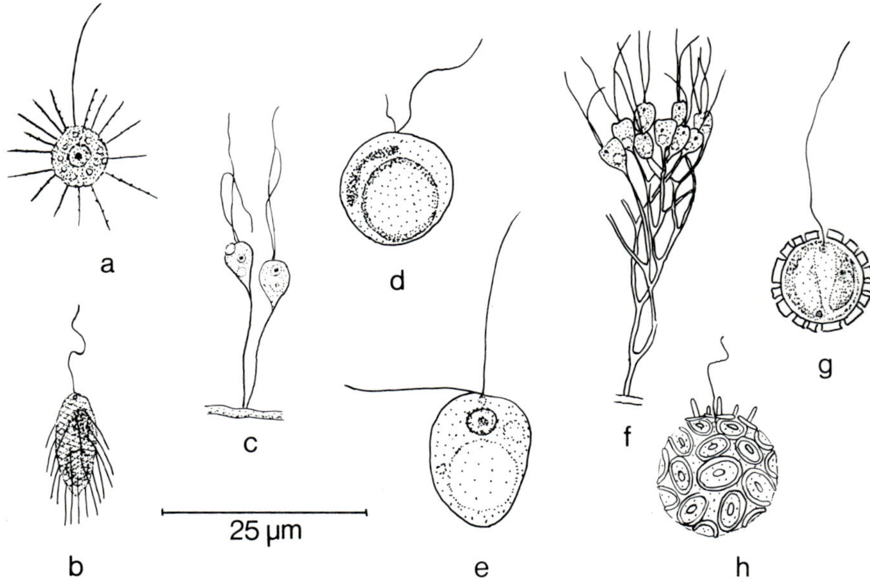

Figure 3. **Chrysomonadina** (Magnification 1100 ×)
a–b **Chromulinidae:**
    a *Actinomonas mirabilis*, Heliozoa-like with radiating pseudopodia; free-swimming or attached by one pseudopodium
    b *Mallomonas dentata*, with silica scales and silica needles; planktonic
c **Isochrysididae:**
    c *Amphimonas globosa*, attached singly by a stalk-like posterior end
d–f **Ochromonadidae:**
    d *Ochromonas granularis*, longish chromatophore pushed to the side by a large round leucosin mass
    e *Monas vulgaris*, without a chromatophore; in putrifying media (faeces)
    f *Dendromonas virgaria*, branched, attached colony
g–h **Coccolithophoridae:**
    g *Pontosphaera heackeli*, transverse section with a coccolith coat, chromatophores inside
    h *Syracosphaera pulchra*, coccoliths spined in parts

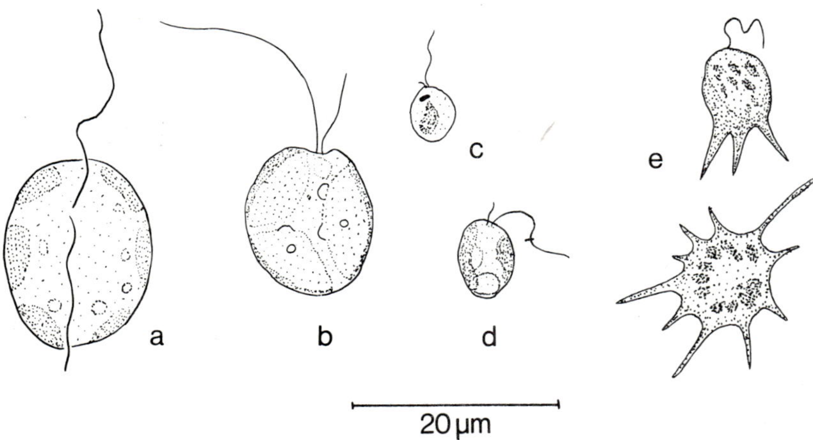

20 µm

Figure 4. Heterochloridina (Magnification 1375 ×)
a–d **Euheterochloridina,** predominantly flagellate:
  a *Olisthodiscus luteus,* with 6 chromatophores
  b *Chlorokardion pleurochloron,* with 3 chromatophores
  c *Chloromeson parva,* 1 chromatophore
  d *Nephrochloris salina,* with 2 chromatophores
e **Rhizochloridina,** aflagellate or incompletely flagellate:
  e *Rhizochloris arachnoides,* form above with one flagellum, form below aflagellate, amoeboid

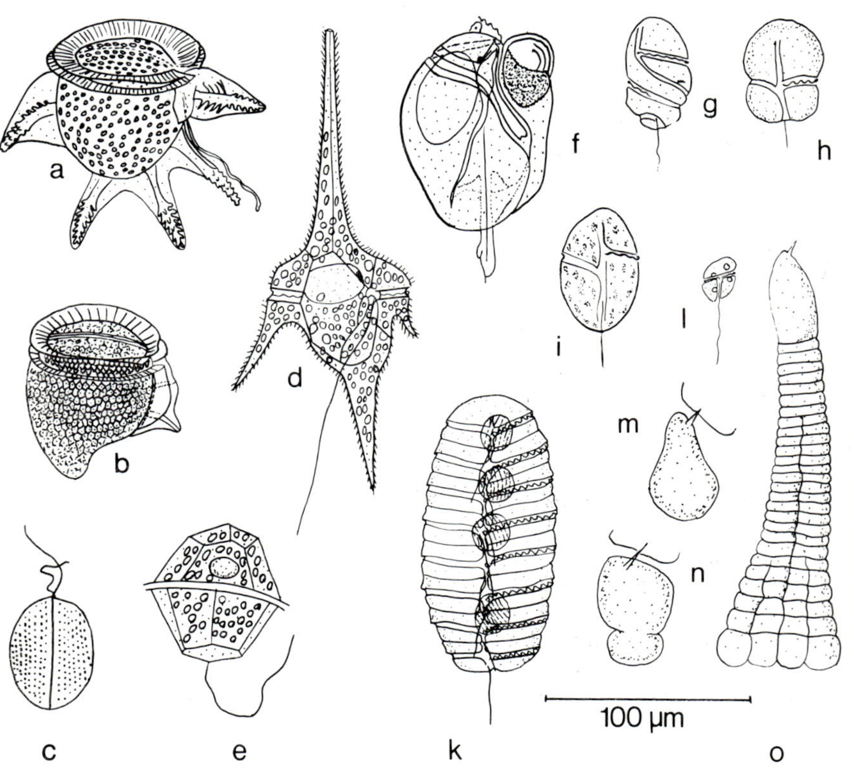
a
b
c
d
e
f
g
h
i
l
m
n
k
o
100 µm

Figure 5. **Dinoflagellata** (Magnification 275 ×)
a–b **Dinophysidae,** with an annular winged ridge at the front and with lateral winged ridges. Pelagic inhabitants of the sea:
  a *Phalacroma jourdani*
  b *Phalacroma mitra*
c **Prorocentridae,** free flagella towards the front, no flagellar groove:
  c *Exuviaella marina*, marine
d–e **Peridinidae,** shells divided up like armour:
  d *Ceratium hirundinella,* armour with horn-like processes which vary with geographical locality and with season of the year; in fresh water and salt water
  e *Goniodoma acuminata,* likewise in fresh water and salt water
f–o **Gymnodinidae,** without strong shells; in fresh water and sea water:
  f *Erythropsis cornuta,* in addition to the flagella, posesses a tentacle directed backwards. At the top right, an eye spot with a hyaline lens structure; marine.
  g *Cochliodinium lebourae,* spiral transverse groove, without chromatophores; marine
  h *Gymnodinium dissimile,* with a straight transverse groove, without chromatophores
  i *Gymnodinium racemosus,* with chromatophores
  k *Polykrikos schwartzi,* marine form with multiple nuclei; flagella and flagellar grooves = colonial individual
  l–o *Haplozoon clymenellae,* parasite in the gut of polychaetes. The Gymnodinidae-type swarmer (**l**, magnification 750 ×) attaches itself to the gut wall with a spike (**m**) and with the loss of the flagellum becomes a trophocyte, from which numerous cells arise by division (**n–o**). The distal cells (sporocytes) become 4-nucleate and detach themselves. The motile swarmers slip out into the surrounding water (**l**)

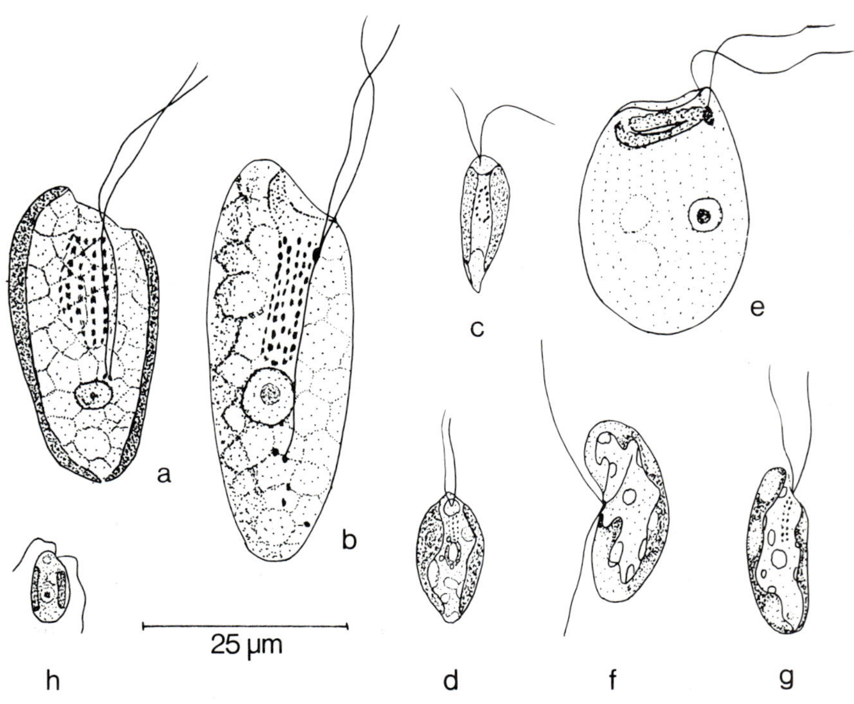

Figure 6. **Cryptomonadina** (Magnification 1375 ×)
  a *Cryptomonas ovata*, very common in fresh water with peripheral greenish chromatophores and typical gullet structure
  b *Chilomonas paramecium*, morphologically similar to *C. ovata* but without chromatophores; often abundant in putrescent marsh water
  c *Rhodomonas pelagica*, marine, with red chromatophore
  d *Rhodomonas lens*, chromatophore red; in fresh water
  e *Cyathomonas truncata*, without a chromatophore; wide gullet formation with a ring of ejectisomes at the front; in standing water
  f *Protochrysis phaeophycearum*, without the deep gullet structure, chromatophores yellow-brown; fresh-water form
  g *Cryptochrysis commutata*, brown to olive-green, with gullet structure
  h *Chrysidella schaudinni*, chromatophores yellowish; lives as a symbiont in the foraminiferan *Peneroplis pertusus*

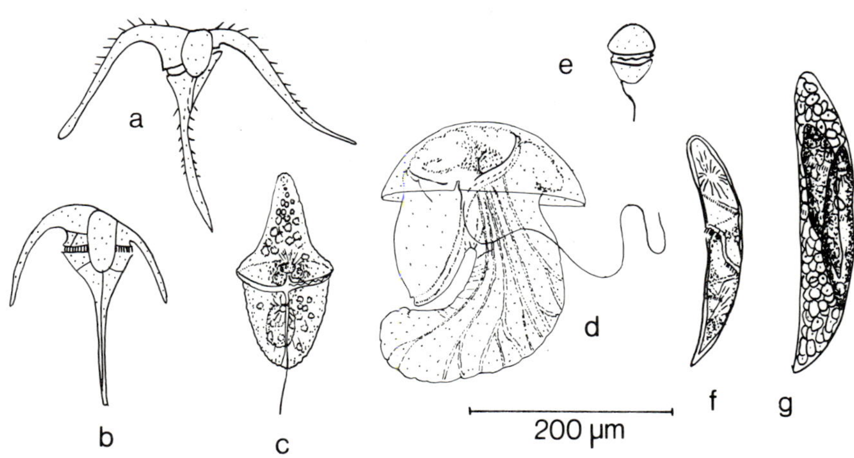

Figure 7. **Dinoflagellata** (Magnification 140 ×)
a–b **Peridinidae:**
  a *Ceratium longipes,* marine
  b *Ceratium tripos,* marine
c–g **Gymnodinidae:**
  c *Gymnodinium dogieli,* marine, with median transverse groove
  d *Pomatodinium impatiens,* marine, with a flat shell sitting on top
e–g *Blastodinium spinulosum,* in the gut lumen of copepods, parasitic, with
yellow-brown chromatophores
  e (magnification 1200 ×) free, flagellate swarmer
  f young aflagellate trophocyte in the gut lumen with spiral grooving
  g older stage in the gut lumen. Repeated divisions (palisporogenesis) lead
  to 2 new trophocytes and numerous sporocytes, which become flagellate
  swarmers **e** after the coat of the mother cell has burst

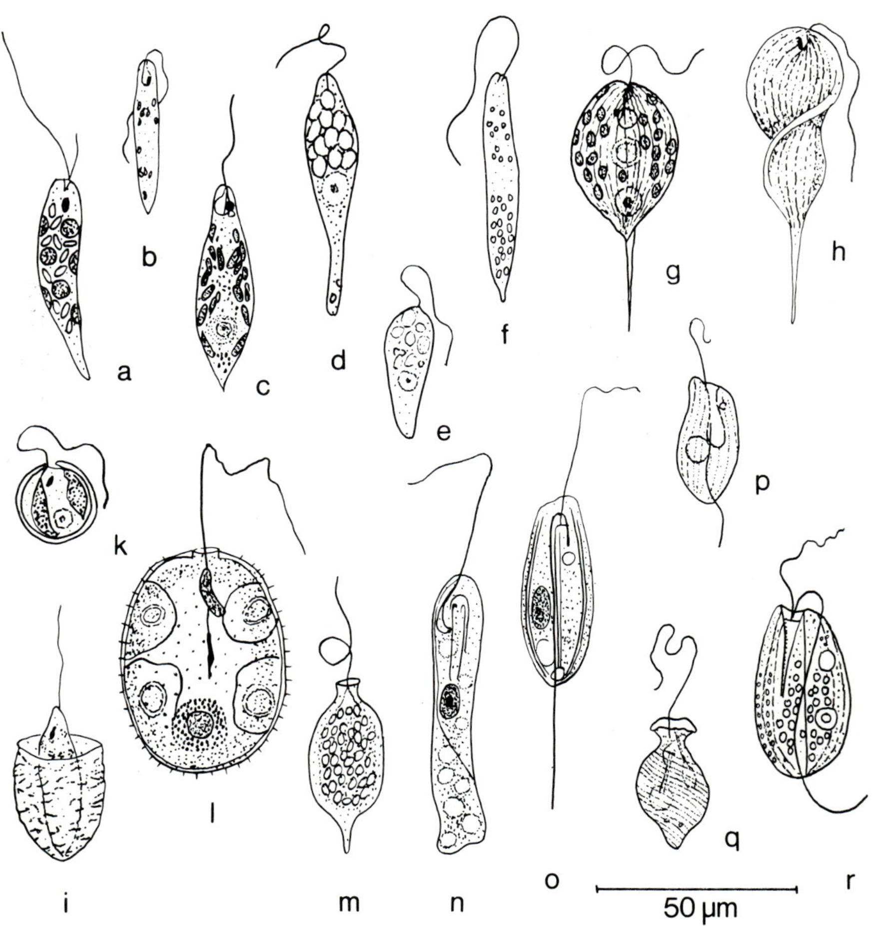
a
b
c
d
e
f
g
h
k
i
l
m
n
o
p
q
r
50 μm

Figure 8. **Euglenoidina** (Magnification 550 ×)
a–m **Euglenidae,** 1–2-flagella, with chromatophores or colourless:
  a *Eutreptia marina,* with 2 unequal flagella; marine
  b *Euglena gracilis,* 1 long flagellum. fusiform chromatophores
  c *Euglena viridis,* rounded off at the front, pointed at the back, longish
chromatophores generally arranged in a star
  d *Astasia klebsi,* without chromatophores and stigma; form similar to
Euglena, spindle shaped, in standing water
  e *Astasia comma,* colourless, usually claviform; in fresh water among
rotting leaves
  f *Astasia longa,* colourless, front end cut off at an angle, back end pointed,
powerful flagellum; in marshy water
  g *Phacus longicaudus.* Through the strengthening of the pellicle, *Phacus*
species are rigid and usually levelled off, with platelet-shaped chromatophores. *P. longicaudus* with a long terminal spine; in ponds and marshes
  h *Phacus tortus,* screw-shaped cell body twisted through about 360°; widespread in fresh water
  i *Klebsiella alligata,* with chromatophores and a stigma like *Euglena,* but
with the posterior end anchored in an open capsule; in standing water
  k *Trachelomonas volvocina. Trachelomonas* species have a strong spreading
brownish shell with a neck-like opening for the emergence of the flagellum.
T. volvocina with numerous chromatophores, stigma usually rod-shaped,
shell smooth, dark, spherical to elliptical; in fresh water
  l *Trachelomonas hispida* with spiny reinforcements to the shell, large
chromatophores; fresh-water form
  m *Trachelomonas urceolata,* smooth capsule, urn-shaped, pointed at the
back, large chromatophores
n–r **Peranemidae,** always lacking chromatophores, food uptake through the
cytopharynx (flagellar sac):
  n *Peranema trichophorum,* a free flagellum at the front, the second flagellum
difficult to see lying against the pellicle as a trailing flagellum. Towards
the front in the cell body a rod-like organelle which strengthens the
gullet; paramylum deposits in the cytoplasm. Frequent in standing water
  o *Anisonema truncatum,* rigid, asymmetrically flattened anteriorly, with a
rod-like organelle. Trailing flagellum more powerful than the free-
swimming flagellum; frequent in standing water
  p *Marsupiogaster striata,* likewise biflagellate, asymmetrical, with a cytostome structure as far as the middle of the cell; fresh-water form
  q *Urceolus cyclostomus,* gullet widened at the front to form a cone, cell
body almost rigid, urn-shaped, uniflagellate; in ponds and bogs
  r *Entosiphon sulcatum,* cell body rigid, ribbed, with deep gullet structure,
biflagellate, trailing flagellum very long

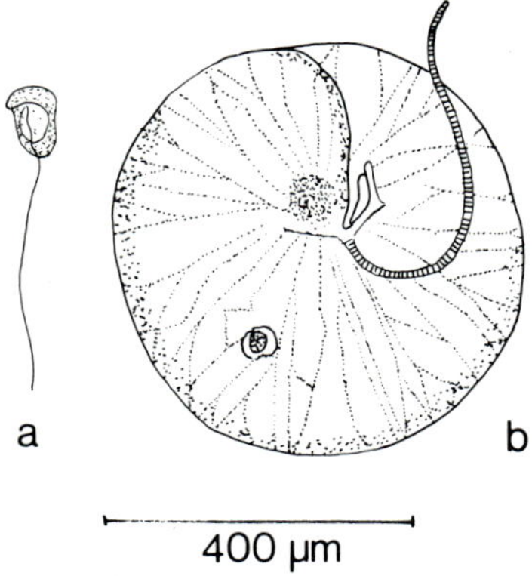

Figure 9. **Dinoflagellata** (Magnification 70 ×)
Gymnodinidae: a–b *Noctiluca miliaris,* marine
   a uniflagellate immature form = swarmer (magnification 275 ×)
   b adult stage, central cytoplasm with a branching reticulum in a gelatinous
sphere with a notched oral cone (peristome) in which there are tooth-like
pegs and a rudimentary flagellum. With ribbon-like cross-striated tentacle.
N. miliaris is the cause of phosphorescence of the sea, particularly
frequent near the coast.

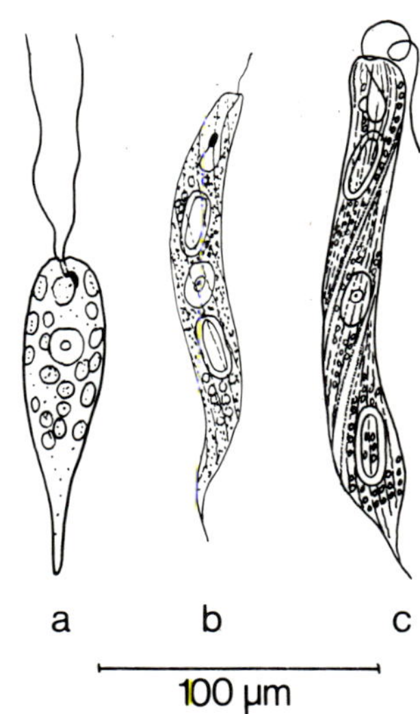

Figure 10. **Euglenoidina** (Magnification 275 ×)
a–c **Euglenidae:**

a *Eutreptia viridis,* green chromatophores disc-shaped, reddish stigma, 2 flagella of equal length; in standing water

b *Euglena spirogyra,* cell body slightly tapered at the front, pointed at the back, flagellum short, chromatophores numerous and disc-like, pellicle yellowish with spiral rows of tubercles, two large paramylum deposits, one behind and one in front of the nucleus; in swamps and ditches

c *Euglena oxyuris,* particularly large species, sometimes up to nearly 0·5 mm long. Pellicle spirally striated, flagellum half the length of the cell body, many discoid chromatophores. Two large paramylum formations, one behind and one in front of the nucleus. In standing water

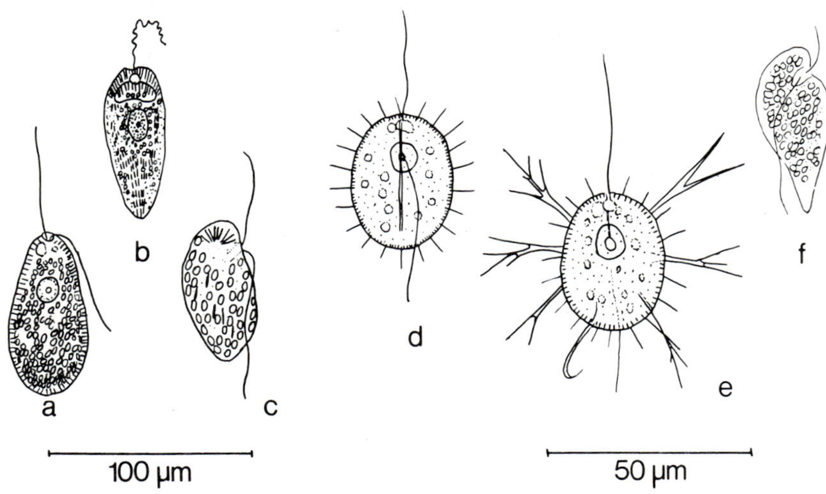

Figure 11. **Chloromonadina.**

a–c Magnification 275 × , d–f Magnification 550 × .

a *Vacuolaria virescens,* swimming flagellum and trailing flagellum almost the same length, numerous chromatophores; lives in ponds and also in mud

b *Gonyostomum semen,* trailing flagellum not visible in the figure; numerous chromatophores with extrusomes in the cytoplasm; living among rotting plants in ponds

c *Merotrichia capitata,* with swimming and trailing flagella; chromatophores, with extrusomes at the front in a fan formation; in water and mud

d *Thaumatomastix setifera,* colourless, truly heterotrophic, cell surface with bristles, with swimming and trailing flagella

e the same species living in mud with free pseudopodia

f *Chattonella subsalsa,* with peripheral chromatophores, ventral depression in the flagellar region (swimming and trailing flagella); in ponds and pools

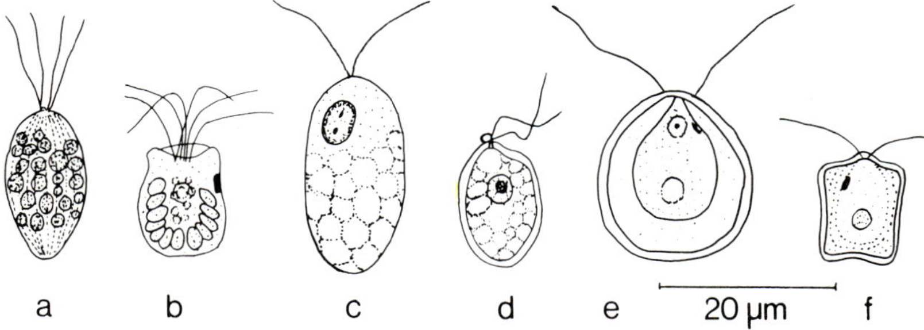

Figure 12. **Phytomonadina** (Magnification 1030 ×)
a–b **Polyblepharididae,** without cellulose membrane:
  a *Polytomella citri*, 4-flagellate, with chromatophores in contrast to most other *Polytomella* species; in fresh water
  b *Pocillomonas flos aquae*, 6-flagellate, with discoid chromatophores; in ponds with clean water
c–f **Chlamydomonadidae,** with cellulose membrane:
  c *Polytoma uvella*, biflagellate, without chromatophores but with starch production, with or without stigma; in putrescent swamp water
  d *Parapolytoma satura*, biflagellate, without a chromatophore, no starch production, no stigma. Front end cut at a slant; in fresh water
  e *Dysmorphococcus variabilis*, biflagellate, cup-shaped chromatophore, with a spreading, often brownish cellulose membrane; in fresh water
  f *Diplostauron pentagonium*, angular membrane, with chromatophore and stigma, biflagellate; in fresh water

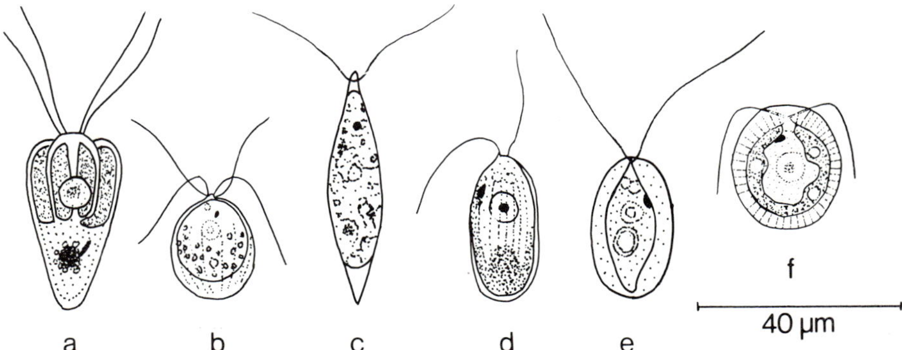

Figure 13. **Phytomonadina** (Magnification 690 ×)
a–b **Polyblepharididae:**
   a *Pyramidomonas tetrarhynchus,* 4-flagellate, with chromatophores, square, with or without stigma; in fresh water
   b *Tetrablepharis multifiliis,* 4-flagellate, ovoid, without chromatophore; in standing water
c–e **Chlamydomonadidae:**
   c *Chlorogonium euchlorum,* biflagellate, elongate, with green chromatophore and red stigma and numerous contractile vacuoles; in standing water
   d *Chlamydomonas steini,* oval with delicate cellulose membrane, chromatophore large, with stigma, often in huge numbers in pools of water and ponds
   e *Sphaerellopsis fluviatilis,* biflagellate, with a gelatinous mass between the cytoplasm and the outer membrane. Chromatophore large, with stigma, 2 contractile vacuoles towards the front; fresh-water species
   f *Haematococcus pluvialis,* biflagellate with a gelatinous zone between the central cytoplasm and the outer cellulose membrane; in this zone are fine cytoplasmic processes. Chromatophore normally green, red when subjected to nutrient deficiencies in pools of water and ponds after rainfall

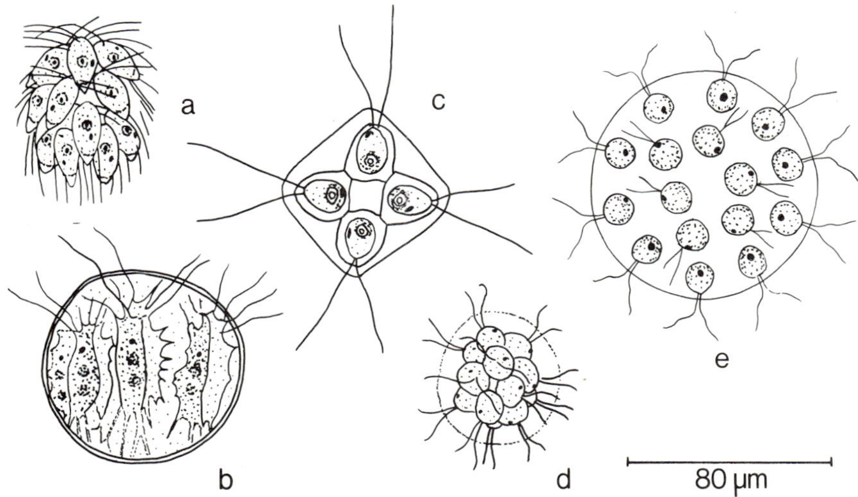

Figure 14. **Phytomonadina** (Magnification 345 ×)
a–e **Volvocidae,** with colony formation, all with green chromatophores and reddish stigma; in fresh water:
  a *Spondylomorum quarternarium,* colony of 15 4-flagellate individuals loosely associated together
  b *Stephanosphaera pluvialis,* colony of 8 elongate biflagellate individuals with branched cytoplasmic offshoots in a gelatinous mass
  c *Gonium sociale,* flat slab-shaped colony with 4 biflagellate cells embedded in gelatine
  d *Pandorina morum,* spherical colony usually with 16 biflagellate, centrally situated cells, enclosed in a gelatinous coat; frequently occurring species
  e *Eudorina elegans,* spherical to elliptical colony usuallly of 32 biflagellate individuals, which are situated peripherally in the gelatinous sphere

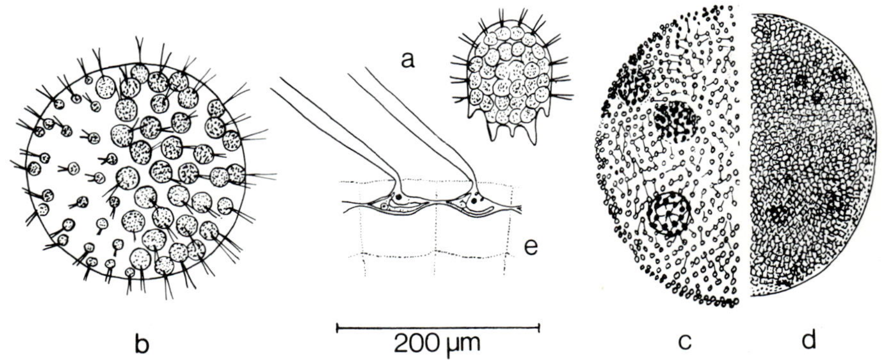

Figure 15. **Phytomonadina** (Magnification 105 ×)
a–e **Volvocidae:**

a *Platydorina caudata,* 32 individuals as a slab-shaped colony, rounded off at the front and with several pointed spurs at the back

b *Pleodorina californica,* 128 cells in a gelatinous sphere, 64 somatic cells on the left which have lost the capacity for division, 64 generative cells on the right, from each of which a daughter colony can arise by division

c *Volvox aureus,* half of a spherical colony with numerous peripherally situated individual forms whose gelatinous coats are at times separated from one another with a hexagonal outline. Predominantly somatic cells, only few generative individuals which are capable of producing daughter colonies

d *Volvox globator,* like *V. aureus* but with even more numerous individual cells (up to 20,000), likewise predominantly somatic cells

e two somatic cells of *V. globator,* highly magnified, biflagellate, with green chromatophore and stigma. The individual cells are connected with one another by cytoplasmic processes.

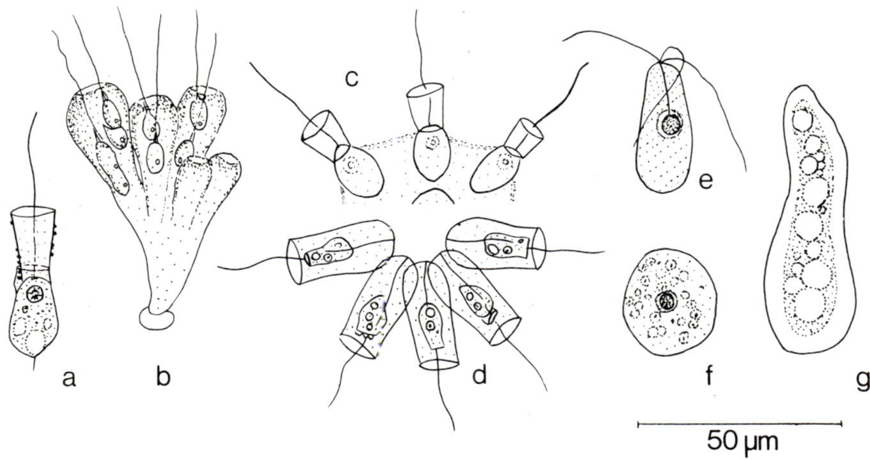

Figure 16. **Protomonadina** (Magnification 550 ×)
a–c **Craspedomonadidae** (Choanoflagellata), uniflagellate with collar; in fresh water:
   a *Codonosiga botrytis,* situated singly on plants and plankton organisms. Nutrient particles are outside, on the rim of the collar
   b *Phalansterium digitatum,* solitary flagellates with narrow collar formation in strong branched gelatinous tubes
   c *Protospongia haeckeli,* part of a free-swimming colony embedded in gelatine
d **Bicoecidae,** biflagellate, with collar:
   d *Bicoeca socialis,* part of a free-swimming colony, flagellates with the trailing flagellum anchored in the matrix; in fresh water
e–g **Tetramitidae** (with division of the nucleolus at mitosis)
e–f *Trimastigamoeba philippinensis:*
   e flagellate stage
   f amoeboid condition with no flagella; in decaying media (human faeces)
   g *Vahlkampfia ranarum,* fresh preparation, no flagellate stage. Parasite of the lower gut of frogs

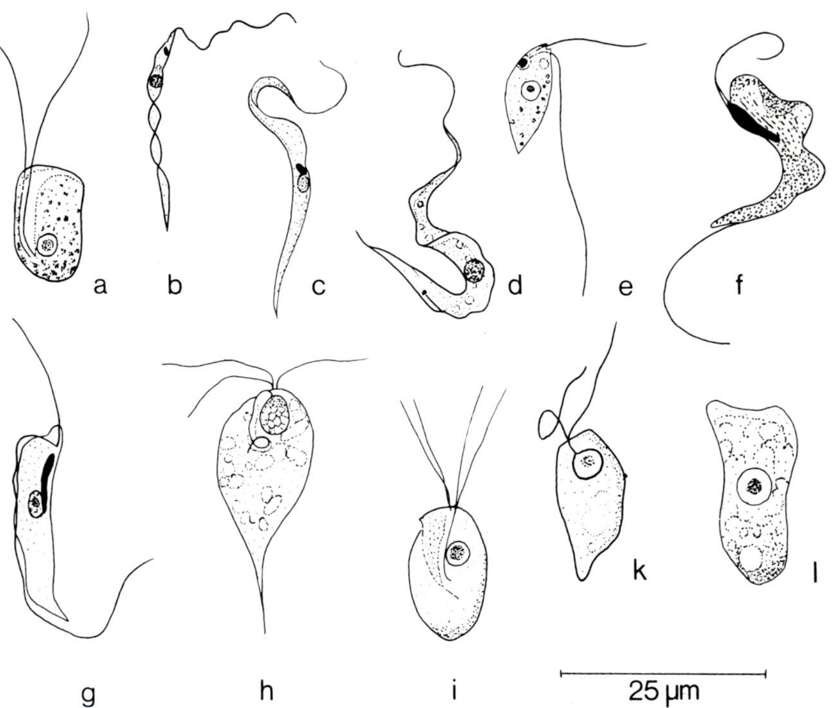
a
b
c
d
e
f
g
h
i
k
l
25 µm

Figure 17. **Protomonadina** (Magnification 1100 ×)
a **Eumonadidae,** heterogeneous Family:
a *Costia necatrix,* with 2 flagella arising from a deep depression, with which the flagellate can adhere by suction to the skin and gills of fish. A pathogen particularly for young fish, in fresh water
b–d **Trypanosomatidae,** with nucleus and kinetoplast; parasitic:
b *Leptomonas ctenocephali,* kinetoplast towards the front = promastigote. In the gut of the dog flea amastigote stages, too, form in the lower gut (compare figure 18 d)
c *Crithidia hyalommae,* kinetoplast near the nucleus = epimastigote. With undulating membrane; a parasite in the body cavity of the tick *Hyalomma aegyptium*; also forms an amastigote stage
d *Trypanosoma theileri.* Kinetoplast towards the back = trypomastigote. With undulating membrane. Blood parasite of cattle, normally apathogenic
e–g **Bodonidae,** with a swimming flagellum and a trailing flagellum and a kinetoplast near the base of the flagella:
e *Bodo saltans,* thickened kinetoplast; in decaying media, coprozoic
f *Trypanoplasma cyprini,* attenuated kinetoplast, with undulating membrane; blood parasite in fresh-water fish, particularly Cyprinidae. Transmission by leeches (alternative host)
g *Cryptobia congeri,* attenuated kinetoplast, trailing flagellum without undulating membrane; intestinal parasite of the sea eel *Conger niger*
h **Retortamonadidae,** with a trailing flagellum in the cytostome:
h *Chilomastix caulleryi,* with three free flagella; intestinal parasite of amphibians, e.g. in frogs and toads
i–l **Tetramitidae,** with peculiar mitosis:
i *Tetramitus rostratus,* with 4 flagella from which an intracellular strand (the rhizoplast) runs to the nucleus. Deep funnel-shaped cytostome formation. Without flagella they can become amoeboid. In decaying media
k–l *Naegleria gruberi :*
k flagellate phase,
1 amoeboid phase, differing from *Vahlkampfia* (figure 16 g) in being able to produce flagella

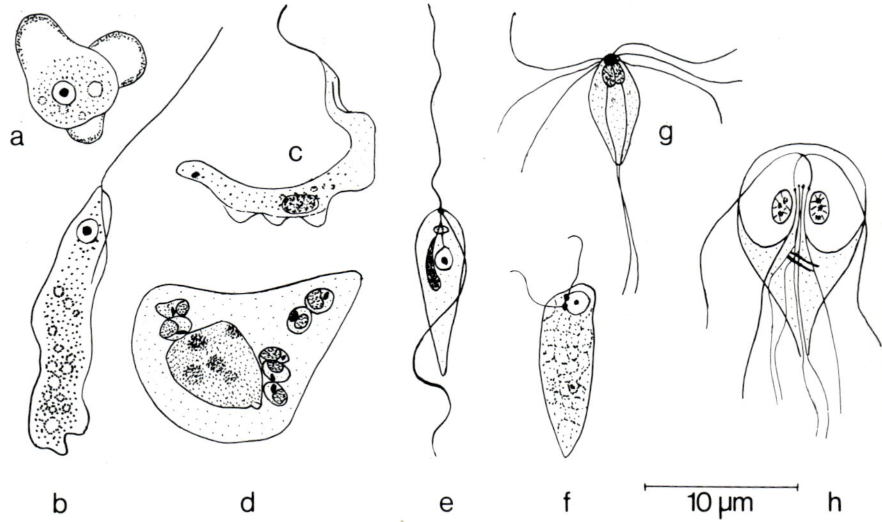

Figure 18. **Protomonadina** (Magnification 2060 ×)
a–b **Cercomonadina,** amoeboid flagellates:
  a–b *Cercomonas bistadialis :*
  a as an amoeba
  b as an amoeboid flagellate with a swimming flagellum and a trailing flagellum; in stagnant waters
c–d **Trypanosomatidae:**
  c *Trypanosoma gambiense.* Kinetoplast towards the rear = trypomastigote. Blood parasite, causal agent of African sleeping sickness in man, transmitted by tsetse flies (*Glossina*)
  d *Leishmania donovani,* seven parasites in one leucocyte, without flagella = amistigote, with only a nucleus and kinetoplast in an oval cytoplasmic body. Causal agent of kala-azar in man, transmitted by sandflies (*Phlebotomus*) in whose gut the flagellate promastigote phase occurs
e **Bodonidae:**
  e *Proteromonas lacertae,* lanceolate to pyriform, swimming flagellum and trailing flagellum; frequent intestinal parasites in lizards
f **Retortamonadidae:**
  f *Retortamonas intestinalis,* with a trailing flagellum in the cytostome and a swimming flagellum; occasional parasite of the large intestine in man
g–h **Distomatidae** = double individuals:
  g *Octomitus muris,* with four pairs of flagella arising from their basal granules, two being trailing flagella. Binucleate. Frequent parasite in the lower small intestine of rats and mice
  h *Giardia intestinalis,* binucleate, with four pairs of flagella. An annular fibril surrounds a large bowl-shaped ventral depression, which serves as a sucker for the attachment of the parasite to the intestinal villi. Parasite of the small intestine of man

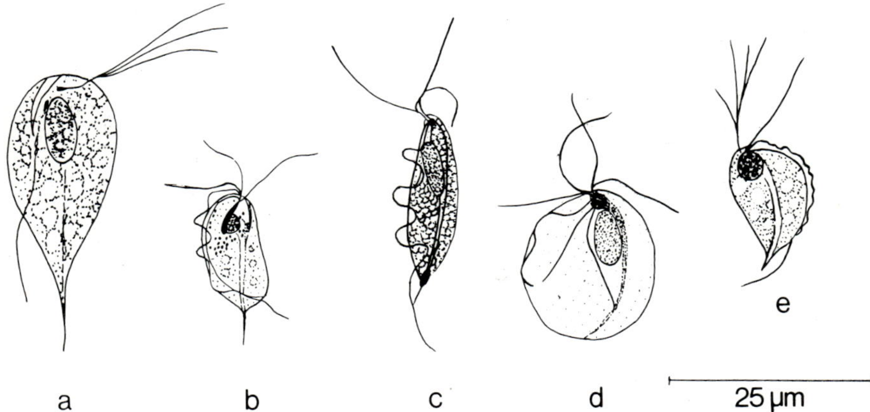

Figure 19. **Polymastigina** (Magnification 1100 ×)
a–e **Trichomonadidae,** with axostyle and 3–5 swimming flagella and one trailing flagellum:

a *Monocercomonas colubrorum,* with long free-trailing flagellum and 3 swimming flagella, an oblong parabasal apparatus, and an axostyle. Frequent intestinal parasite in snakes and lizards

b *Tritrichomonas caviae,* 3 free-swimming flagella, the trailing flagellum forms an undulating membrane, a rod-shaped parabasal apparatus at the basal body region of the flagella. An axostyle. Intestinal flagellate in the guinea pig

c *Tritrichomonas foetus,* with 3 swimming flagella and an undulating membrane formed by the trailing flagellum. An axostyle. Genital parasite of cattle involving the foetus, causing inflammation and abortion. Transmission by coitus

d *Trichomonas vaginalis,* with 4 free flagella and a short undulating membrane. Urogenital parasite in man. Can produce inflammation of the vagina and discharge: venereal disease

e *Pentatrichomonas ardin delteili,* with 5 swimming flagella and a longer undulating membrane. Parasite of the large intestine in man, apathogenic

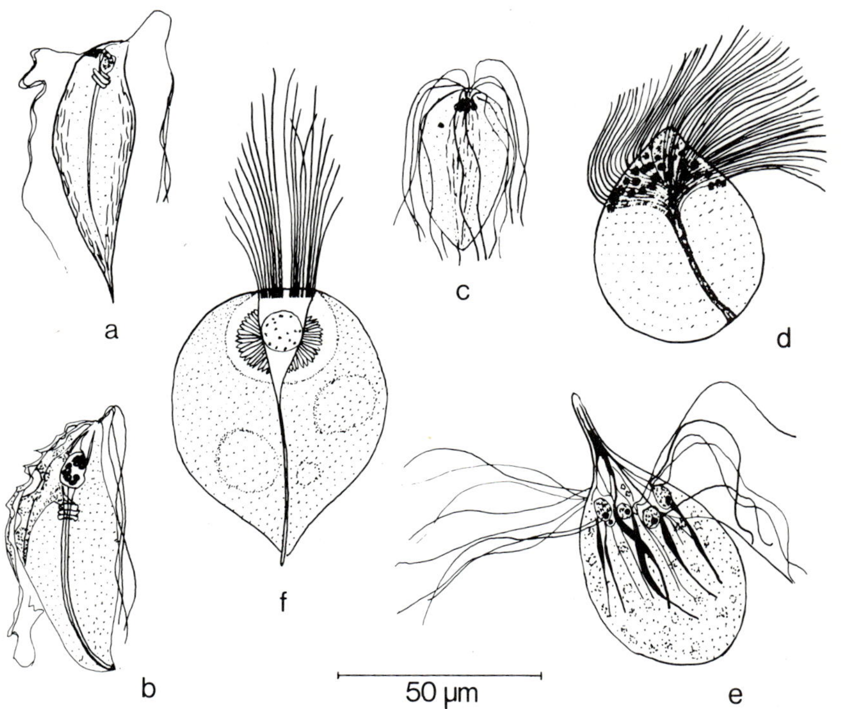
a
b
c
d
e
f
50 µm

Figure 20. **Polymastigina** (Magnification 550 ×)
a–b **Trichomonadidae**:

a *Devescovina striata*, like *Tritrichomonas* with 3 swimming flagella and a trailing flagellum which is wide like a ribbon but without the formation of an undulating membrane. The basal fibril becomes a flat triangular costa (towards the front, to the left of the nucleus). At the front, the parabasal body forms a spiral around the axostyle. Intestinal parasite in xylophagous termites.

b *Macrotrichomonas pulchra*, like *Devescovina*, but with a more strongly developed costa. Widely distributed in termites of the genus *Glyptotermes*

c–d **Calonymphidae**, with multiplication of the organelles:

c *Coronympha octonaria*, colonial individual equivalent to an eightfold multiplication of *Devescovina*, with the parabasal bodies grouped together to form a rod on the anterior crown of the 8 nuclei. Intestinal parasite in termites of the genus *Kalotermes*

d *Calonympha grassii*, further multiplication of the nuclei, the swimming flagella with their basal bodies and the parabasal body, but without trailing flagellum and costa. The axostyles are bundled together. Intestinal parasite of *Cryptotermes grassii*

e **Pyrsonymphidae**, without parabasal body, with intranuclear central spindle:

e *Microrhopalodina multinucleata*, colonial individual with various numbers of nuclei and a corresponding number of axostyles, which reach forwards in a proboscis by which the parasite can attach itself in the gut of its host. All flagella equal, as in *Calonympha*, with 4 flagella to each axostyle. Parasite of *Cryptotermes dudleyi*

f **Hypermastigidae**, uninucleate with numerous flagella and axostyles:

f *Lophomonas blattarum*, only with equal flagella, as in *Calonympha*; their basal bodies are arranged in a circle. Axostyles united into a bundle which splits near the front around the nucleus and surrounds it with the parabasal bodies which together form the parabasal apparatus. Intestinal parasite of the domestic cockroach *Blatta orientalis*

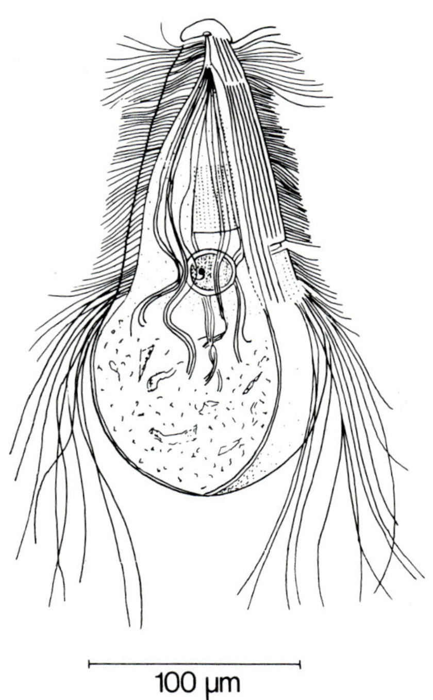

Figure 21 **Polymastigina** (Magnification 275 ×)
Hypermastigidae: *Trichonympha collaris*, with numerous flagella of various lengths, only one nucleus. The parabasal bodies are prominent as thread-like structures around the nucleus. Cell body organized in zones: at the front, the dome-shaped operculum, enclosing the elongate rostrum, which is divided from the posterior section of the cell body by a circular constriction. No axostyle. The ectoplasm is grooved in longitudinal lamellae in the region of the flagella. The flagella originate at the bottom of the resultant longitudinal ribs from basal bodies which are arranged in longitudinal rows. Intestinal parasite of the termite *Termopsis angusticollis*

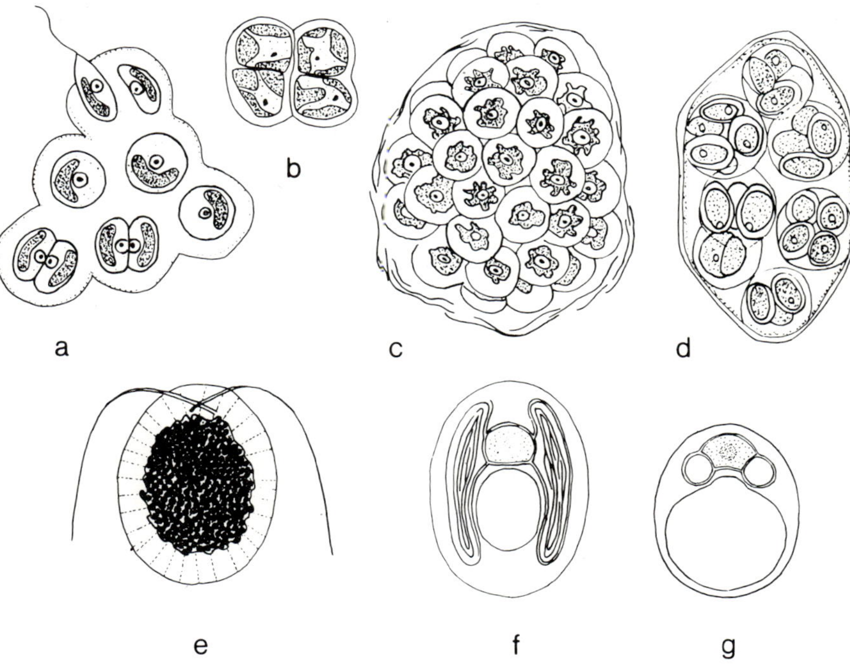

Figure 22 a–d. **Palmella stages**:
   a *Chromulina* (Chrysomonadina) see figure 2 a, b. Partly in the division phase. Top left; development back into the flagellate form
   b *Euglena gracilis* (Euglenoidina) see figure 8 b
   c *Haematococcus pluvialis* (Phytomonadina) see figure 13 f
   d *Chlamydomonas brauni* (Phytomonadina) see figures 13 d and 24 d
e–g **Changes in the chromatophores**:
   e *Haematococcus pluvialis*, red haematochrome pigmentation (compare green normal phase, figure 13 f)
f, g *Ochromonas danica*,
   f oblong chromatophores, either side of the nucleus in pigmented flagellate
   g proplastids reduced to colourless round vesicles after culture in the dark

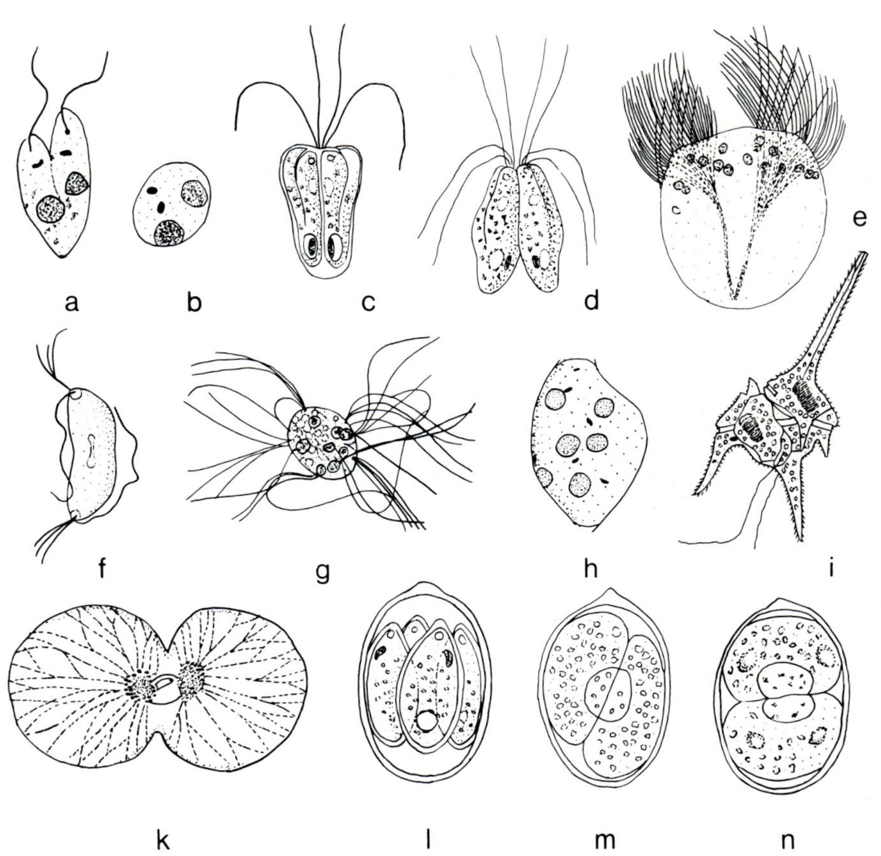

Figure 23. **Processes of division in flagellates:**
a, b binary fission of *Leptomonas ctenocephali* (see figure 17 b):
a promastigote form,
b amastigote stage
c, d binary fission of *Pyramidomonas tetrarhynchus* (see figure 13 a)
e binary fission (plasmotomy) of *Calonympha grassii* (see figure 20 d)
f, g division of *Monocercomonas colubrorum* (see figure 19 a) after dispersal
of the axostyle:
f binary fission
g multiple division (schizogony)
h *Crithidia hyalommae* (see figure 17 c) multiple division of the amastigote
form.
i binary fission of *Ceratium hirundinella* (see figure 5 d)
k *Noctiluca miliaris*, binary fission
l division cyst of *Chlamydomonas angulosa*
m, n division cyst of *Chlamydomonas longistigma* (see figure 22 d)

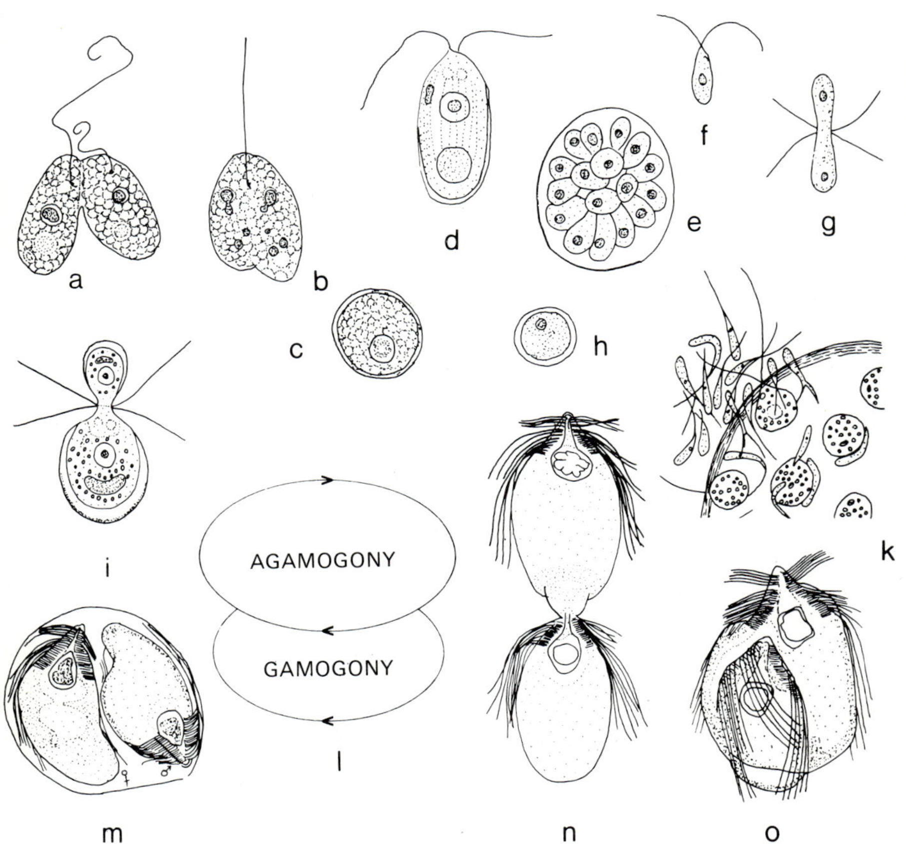
a
b
c
d
e
f
g
h
i
AGAMOGONY
GAMOGONY
l
m
n
o
k

Figure 24. **Sexuality of the flagellates.**
a–c *Scytomonas pusilla*:
  a, b fusion of two gametes: isogamy (hologamy)
  c zygote
d–h *Chlamydomonas steini*:
  d normal vegetative form
  e encystment with internal gamete formation
  f free gamete
  g fusion of two gametes of equal size: isogamy
  h zygote
  i *Chlamydomonas brauni*, fusion of two morphologically unequal gametes: anisogamy
  k *Eudorina elegans.* Microgametes swarm round a colony with macrogametes (only a part shown) and fusion anisogamy (oogamy) occurs.
  l diagram: reproduction of the flagellates most commonly by agamogony. Species with sexuality occasionally also undergo a gamogony cycle
m–o *Trichonympha* sp.
  m encystment with gamete formation
  n–o fusion: anisogamy.

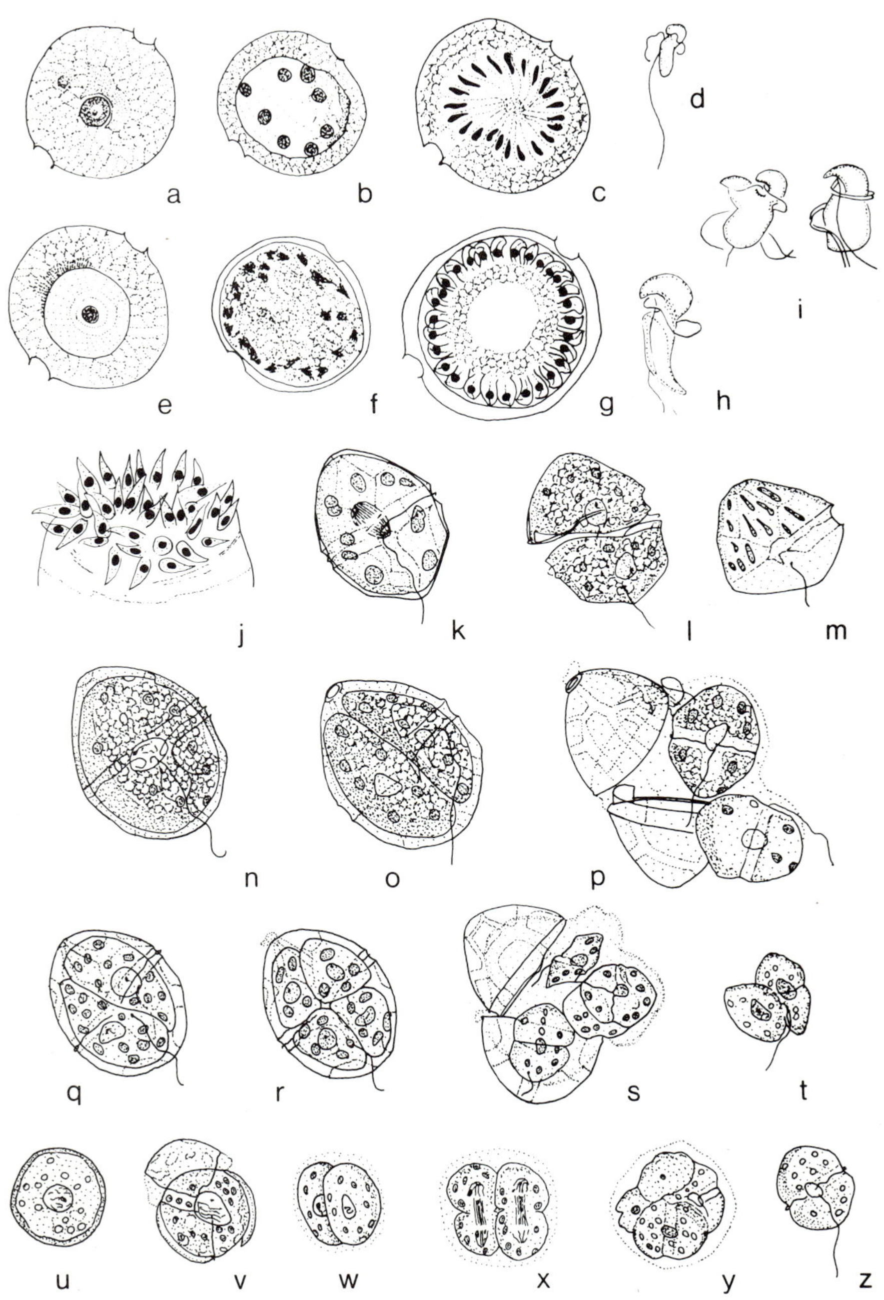

Figure 25. **Sexuality in the dinoflagellates.**
a–h *Coccidinium duboscqui*
  a–d development of microswarmers:
  a uninucleate parasite
  b schizogony
  c formation of swarmers
  d free microswarmer
  e–h development of macroswarmers:
  e uninucleate parasite with abundant nutrient
  f schizogony
  g peripheral arrangement of the swarmers
  h free macroswarmer
i *Coccidinium mesnili*, fusion of the gamete-swarmers
j *Noctiluca miliaris*, multiple division, formation of swarmers (see figure 9 a)
on a circular segment of the surface of the cell body
k–z *Glenodinium lubiniensiforme*
k–p agamogony:
  k normal trophozoite
  l agamous binary fission
  m daughter cell with onset of production of the second half of the cell body
  n large trophozoite
  o internal agamous division of the large trophozoite (swarmer formation)
  p release of both agamous swarmers
q–z gamogony with subsequent agamous swarmer formation:
q–r formation of 4 gamete-swarmers through internal divisions
  s release of gamete-swarmers
  t fusion
  u zygote
  v rupture of the zygote sheath by the production of jelly
  w–x agamous divisions with the production of 4 daughter cells
  y degeneration of 2 daughter cells and formation of 2 agamous swarmers
  z free agamous swarmer = young trophozoite

CLASS II: RHIZOPODA (VON SIEBOLD 1845)
ROOT-FOOTED ORGANISMS

THE RHIZOPODA HAVE NO PERMANENT ORGANELLES FOR LOCO-
motion. Pseudopodia which originate from the cell plasma from
time to time for this purpose give the Class its name, since one
of the possible types of pseudopodium is the **rhizopodium**.
These are finely branched, like roots, often filiform, and can
fuse together again with one another to form an **anastomosing**
reticulum (figure 42 a, c). The subdivision of the Class Rhizopoda into Orders is made with reference to the presence or
absence of internal or external skeletal elements of the body of
protoplasm. Since the Foraminifera are clearly the most highly
specialized group, the Rhizopoda furnished with a shell will here
be treated last.

Order 1: Amoebina (Cash and Hopkinson 1905, **Amoebaea**
Ehrenberg 1830), figures 26–30

The amoebae ("changing animalcules") are naked bodies of
protoplasm, without skeletal elements, which can change their
shape within certain limits by protoplasmic streaming and
pseudopodia formation. They are heterotrophic and therefore
have a purely animal holozoic system of nutrition. As a result
of varying consistency of the cytoplasm, wider blunt **lobopodia**
(e.g. figure 26) or thinner, often tapering **filopodia** (figure 27 c,
d) are formed as pseudopodia. A peripheral stouter and more
hyaline ectoplasm can usually be distinguished from a more

strongly vacuolated or granulated endoplasm. In the endoplasm are situated one or more nuclei, the food vacuoles and, in fresh-water and soil amoebae, contractile vacuoles. The nuclear structure and the type and means of locomotion of the pseudopodia serve as criteria for further systematic subdivision.

The **Rhizomastigina** which can produce flagella have already been mentioned in the appropriate place in the Flagellata (see figures 2b, 4e, 16e, 17k,l, 18a,b). The genus **Vahlkampfia** (figures 16g, 28a), which is assigned to the **Tetramitidae** on account of the nature of its mitoses, is the only one which always remains aflagellate. A careful study of mitosis is therefore necessary to distinguish it from the genus **Hartmanella** (also **Acanthamoeba**) which belongs to the **Amoebidae** (figure 29a). All of these small amoebae generally move by forming a broad lobopodium, and are often also simply lumped together, without further differentiation, as **Limax amoebae**. A more extensive formation of lobopodia, with vigorous cytoplasmic streaming and changing of form, occurs in the substantially larger **Chaidae** (figure 26a–c), which includes the well-known species *Chaos diffluens* (= *Amoeba proteus*). *Amoeba gorgonia* (figure 27b), which occasionally uses a pseudopodium as a stalk for attachment, is taxonomically close to the genus **Chaos**. The genus **Pelomyxa** (figure 27a), which is always multinucleate, includes amoebae with sluggish movement. The viscous cytoplasm has a margin of ectoplasm from which the lobopodia grow as undulating or hernia-like bulges. These amoebae are oval in shape. The ectoplasmic zone of the **Thecamoebidae** (figure 27e) is even stronger and as thick as a membrane. In their locomotion and formation of folds, the Thecamoebidae show deformation of the anterior sack-like ectoplasm without production of true pseudopodia. They include the binucleate *Sappinia diploidea* (figure 28b), first isolated from lizard faeces; this species is of biological interest with regard to its sexuality. The generally very hyaline filopodia of the **Mayorellidae** (figures 27c, d, 28c) are different from lobopodia. They do not determine the forward movement of the amoebae significantly, but function predominantly in uptake of nutrients.

Many **Amoebidae** live parasitically in regions of gut colonized by bacteria. The main criteria for their systematic classification are the structure of the stained nucleus; nuclear material may be compact (figure 29 b, c), diffuse (figure 29 d) or, in the important genus **Entamoeba**, annularly arranged with an enclosed nucleolus (figures 28 e, f, 29 e, f). The family includes the causal agent of amoebic dysentery (figures 1 e, f, 29 e). Parasites which suck out the cytoplasm of algae form the genus **Vampyrella** (figure 28 d). Others of biological interest are the **Paramoebidae** which contain an extra organelle, able to undergo division in addition to the nucleus (figure 28 h), and the **Acrasiae** (figure 28 g) which sometimes unite to form complex pseudoplasmodia (figure 155). Finally the **Piroplasmidae**, parasites of red blood corpuscles, should also be mentioned here; in contrast to the malaria parasites, which also live in erythrocytes, the Piroplasmidae have no sexuality and hence no sporogony, but reproduce only by binary fission or multiple division in the vertebrate host and in the carrier hosts, ticks (figure 30).

Because they infect red blood corpuscles, and despite the absence of sporogony, the Piroplasmidae are often classified with the Coccidia, more specifically with the malaria parasites. Electron-microscopic studies show structures, including the paired organelles (P) and conoid (C) shown in figure 140, which, together with other fine structures, are now designated the apical complex. Protozoa with an apical complex are combined as the **Apicomplexa**. Since the Piroplasmidae have the apical complex typical of Coccidia, the possibility of combining them with the Coccidia again arises. The main outstanding question, however, is whether this conformity is really homologous and ontogenetic, or is with regard simply to the penetration by the Piroplasmidae into erythrocytes as host cells merely analogous, i.e. behavioural; in this case, the lack of sporogony in the Piroplasmidae would be understandable.

Order 2: Heliozoa (Haeckel 1866), figures 31–33

The Heliozoa, or "sun animalcules", have a spherical body from which very thin pseudopodia radiate out on all sides like rays; these have a firm axis and are called **axopodia**. Two zones of cytoplasm can frequently be recognized. When the ectoplasm is coarsely vacuolated, it is distinguished as the cortical layer from the more finely vacuolated inner medullary substance (figure 32 a). The cortical layer contains the contractile vacuoles, the medullary substance of the nucleus or nuclei. Several Heliozoa are classified together as the suborder **Actino-phrydia** (figures 32 a, 33 a) in which the axial threads of the axopodia arise at the base of the cortical substance or at the nucleus, while in the larger group, the **Centrohelidia**, the axial threads emanate within the medullary substance from a special central granule (figures 32 b, 33 b–g). In this suborder there is frequently strengthening of the cell body with a gelatinous coat as well as with foreign (xenogenous) or self-engendered (autogenous) inclusions. The **Helioflagellidae** (figure 31) are transitional forms (with the Flagellata). They have true axopodia but these arise at a centrosome which acts as a central granule. Some flagellates without axopodia, but with very fine radiating filopodia (figure 3 a), also give the impression of being like the Heliozoa. According to this distinction, the genus **Clathrulina** (figure 33 h) no longer belongs to the Heliozoa and, with its fine filopodia and perforated lattice-work sphere as an exoskeleton, is assigned to the Testacea. With few exceptions (figure 32 b), the Heliozoa are free-swimming Protozoa, living mainly in fresh water.

Order 3: Radiolaria (Müller 1858), figures 34–35

The Radiolaria, or "ray animalcules", like the Heliozoa, generally have a spherical basic form from which axopodia or fine filopodia, as well as rhizopodia, radiate outwards. They are pelagic inhabitants of the warmer seas. In the true Radiolaria, the endoplasm, which contains the nucleus, is always sur-rounded by a strong membrane furnished with pores which is called the **central capsule.** The **Peripylea** or **Spumellaria** have a central capsule which is penetrated all round by numerous pores (figure 34 a), the **Monopylea** or **Nassellaria** a central capsule with one circumscribed pore plate (figure 34 b), the **Tripylea** or **Phaeodaria** a capsule with an astropyle which is opposite two parapyles (figure 34 c). The extracapsular protoplasm is a branching reticulum and usually forms a thick bubbly structure, with numerous lacunae filled mainly with a gelatinous substance (**calymma**). In the Phaeodaria, a yellow-brown mass of pigment, the **phaeodium**, is also situated here, in front of the astropyle. The deposition or formation of strong skeletal elements, already met with in the Heliozoa, is usually much more highly developed in the Radiolaria. Inclusions of foreign bodies such as diatoms occur only occasionally here. The autogenous skeletal elements consist of amorphous silicic acid (opal). They can lead to most beautiful and complex skeletal formations, which, on account of their durability, in some places form a considerable proportion of the sea bed. Not all Radiolaria have these inorganic strengthening elements (figure 34 d, m). In addition to radially arranged silica needles (figure 34 h), concentric skeletal structures also occur (figure 34 e, i); the lattice-like stromata (figure 34 l), which can be spherical, galeate, cage-like, urn-shaped or campanulate, are examples of the latter. When several of these lattice skeletons are produced in a range of sizes, they are connected to one another by means of radial structures (figure 34 k). Silica needles applied to the outside (figure 34 k) or long feeler-like processes (figure 34 f) increase the ability of the Radiolaria to float. Irregular root-like branching also occurs (figure 34 g). The Radiolaria are given their colour, and often made very opaque, by the opal lustre of the silica skeleton, by pigments and oil globules embedded in the cytoplasm, particularly in the central capsule, by nutrient materials which have been taken up, by metabolic products, and often also by pigmented symbionts (**Zooxanthellae**) which live in the cytoplasm outside the central capsule, and by reddish, green or brown central capsule itself.

The **Actipylea** or **Acantharia** are distinct from the true Radiolaria. A central capsule is by no means always present (figure 35 a). When a central capsule is evident (figure 35 b), it is relatively large, and comprises a thin elastic membrane without pores; the membrane is pierced by at most 20 radially arranged skeletal needles formed from strontium sulphate. These needles are united at the centre or else cross each other there. The gelatinous substance of the Acantharia is not enclosed in the ectoplasm, as it is in the true Radiolaria, but is formed as a peripheral covering. Contractile **myonemes** at the outer end of the radial skeletal needles can cause changes in the cell circumference (figure 35 b). Additional concentric lattice and plate-like structures can be present in the Acantharia too, zooxanthellae live within the capsule here, in contrast to their position in the Radiolaria. All of these differences support the classification of the Acantharia as another separate group of Protozoa between the Heliozoa and Radiolaria. Since some species in fact have a central capsule, they can be placed as transitional forms at the beginning of the true Radiolaria.

Order 4: Testacea (Schultze 1854), figures 36–38

The Testacea, also called Thecamoebae (not to be confused with the genus Thecamoeba of the Amoebina) are, as the name implies, amoebae with a shell or test. They differ from the naked amoebae only in the formation of the test; there are no internal skeletal elements. The shell, which is always unilocular (**monothalamous**), is a gelatinous, pseudochitinous or keratinous substance secreted by the cytoplasm. It can be reinforced in some species by the addition of foreign bodies such as sand, quartz, diatom shells and other plant or animal skeletal remains, or by autogenous secretion of silica; in such cases the shells become opaque. The shells of the Testacea possess only one aperture (**pseudostome**) for the emergence of the pseudopodia, which are lobopodia or filopodia. Accordingly, familial relationships with the corresponding Amoebina are assumed. The inner cytoplasm is often subdivided into a nuclear zone at the posterior end, a middle zone with food vacuoles and the contractile vacuole, and an anterior zone, with few inclusions, which is the area for the formation of the pseudopodia. The Testacea are nearly all fresh-water organisms and live under water on plants and on the soil, on moors, and in damp moss. The classification is based on the structure either of the shells or of the pseudopodia but, since frequently only the shells are found, their characteristics will be considered here as the basic criterion.

The **Arcellidae** (figures 37 a–d, 38 a–b) have an organic shell structure which is usually strong, sculptured, and yellowish or brownish in colour; generally they have slender lobopodia and occasionally also filopodia. The **Difflugiidae** (figures 36 a, b, 37 e, f) have a shell strengthened by foreign bodies (**xenosomes**) and also produce lobopodia or filopodia. The **Euglyphidae** (figures 37 g–k, 38 c) reinforce their shells with autogenous deposits (**idiosomes**) which consist of silicic acid, and whose form is characteristic for individual genera. For example, they are elliptical to round in *Nebela*, four-cornered in *Quadrula*, roof-tile-shaped platelets in *Euglypha*, and rod-shaped, twisted, hexagonal or irregular in *Lesquereusia*. The pseudopodia are once again lobopodia and filopodia. *Clathrulina* (figures 33 h) lies outside this classification, with a silicic acid latticed sphere reminiscent of the Radiolaria. Only the genus *Paraquadrula*, which belongs to the Euglyphidae and which resembles the genus *Quadrula* in appearance (figure 38 c), forms autogenous limestone platelets instead of silicic acid and thus resembles the Foraminifera.

Order 5: Foraminifera (d'Orbigny 1826), figures 39–42

The Foraminifera—like the Testacea—are Rhizopoda with shells. Whereas the shells of the Testacea are always unilocular (monothalamous), only the more primitive Foraminifera are still monothalamous. The majority are **polythalamous** and multilocular. The partitions between the chambers are then linked to one another by one or more openings (**foramina**) (figure 43 n, o). "Foraminifera" therefore means "perforated-shelled". The Foraminifera are fossil or recent marine Protozoa, with the exception of the Allogromiidae which live predominantly in fresh water. They live almost exclusively as **vagile** (motile) or **sessile** (attached) forms on the sea bed and on algae, and therefore belong to the **benthos**, some of them occurring at depths of over 1000 metres. Only two families (with about 25 species) are **pelagic**, living floating in the water (**plankton**). Since more than 300 genera have been described, this represents a tiny proportion. The shells of the Foraminifera can be perforated to the outside (**perforata**) or they may lack these pores (**imperforata**). The size of the pores, through which the pore canals of the shell emerge, varies in the different species between 0·5–15 µm. The pseudopodia of the Foraminifera are always rhizopodia; they are also called **reticulopodia**. They pass through a special opening (**mouth** or **aperture**) to the outside or, in perforated shells, through the pores too, if these are sufficiently large. Form and structure of the shells can be very diverse. The **Allogromiidae** (figure 42 a) have a chitinous casing. These are predominantly fresh-water organisms and have for this reason generally been classed with the Testacea. However, they exhibit important characteristics of the Foraminifera in the formation of their rhizopodia and gametes. They have no pores in the shells. The **Agglutinantia** have sticky shells on which can be found quartz grains, calcareous sand, diatom shells, sponge needles, etc., all cemented together with a cement produced by the protoplasm which contains calcium carbonate or iron hydroxide. Here, too, the shells are nearly always without pores. The more highly developed Foraminifera have shells in which, essentially, calcium carbonate is laid down on a thin organic foundation. These shells are made of crystalline calcareous spar and polarize light. Calcareous shells such as these, without pores and with only a mouth for the rhizopodia, constitute the group **Calcarea imperforata**, while the **Calcarea perforata** have perforations. The shells of the Foraminifera, which in part have additional spines, ridges, etc., tend to persist and constitute considerable deposits of oceanic sediment.

A *Catalogue of Foraminifera* published in New York comprises over 50 volumes. With such a plethora of forms, our reference to the taxonomy can only give an extremely fragmentary selection:

the **Allogromiidae** (figure 42 a) have already been mentioned.
Rhabdamminidae (figure 39 b): shells gelatinous or sticky, unilocular, imperforate, not coiled.
Ammodiscidae (figures 39 a, 40 a): shells likewise monothalamous, agglutinated or even calcareous, imperforate, but wound in a spiral.
Nodosinellidae (figure 39 c): shell covered with sand or calcareous, usually imperforate, but polythalamous in a linear arrangement.
Miliolinidae (figure 41 a): shell polythalamous, imperforate, usually completely calcareous, more rarely sticky. Each of the spirally arranged chambers generally spans half a circumference of the shell.
Peneroplidae (figures 40 b, 41 b): polythalamous, calcareous, chambers at first in a flat spiral, then in a more rectilinear arrangement. Only the first chamber (**proloculus**) is perforated.
Textularidae (figure 42 b): shell sandy to calcareous, usually perforate, chambers arranged in alternating rows.
Trochamminidae (figure 41 c): polythalamous, chambers spirally arranged in a single row about an axis, with adhesive

sand or completely calcareous, imperforate.

Rotalidae (figures 40 c, 41 d–f, 42 c): shell polythalamous, always calcareous, perforate in a flat spiral or in a pyramidal trochospiral (turbospiral), with the first chamber (proloculus) as the centre.

The size of Foraminifera usually lies between 20 µm and 1 mm, although there are also many larger species. The present-day species *Psammonyx vulcanicus* (figure 39 a) can reach 6 cm and the fossil species *Nummulites gizehensis* was 11–12 cm with a shell thickness of 1 cm.

Reproduction and Sexuality in the Rhizopoda

The normal agamous **reproduction** of the **Amoebina** is binary fission which begins, as in the Flagellata, with the mitotic division of the nucleus. Since the shape of the Amoebina is constantly changing, there is no set direction for the division of the cytoplasm (figure 43 a–c). The two daughter nuclei divide themselves between the daughter animals which are formed. For the multinucleate *Pelomyxa* (see figure 27 a) the possibility of plasmotomy, the constricting off of multinucleate daughter amoebae, has been discussed. Multiple divisions of amoebae will be considered in the section on cyst formation (figure 116). The course of agamous division in the **Heliozoa** (figure 43 d) is the same as that in the Amoebina. In uninucleata species the mitotic nuclear division again takes place first, together with the division of the central grain when this is present. In multinucleate forms, the nuclei present may be divided among the separating daughter animals and their numbers completed by mitosis during a further course of plasmotomy. During binary fission of the **Radiolaria**, division of the central capsule occurs after the division of the nucleus (figure 43 e). Further development proceeds differently according to species. In simple forms, as in the case of *Aulacantha* (figure 34 h), the division of the extracapsular cytoplasm is accompanied by a distribution of the skeletal needles among the daughter cells, which subsequently complete their skeletons. If the skeleton cannot be divided, one daughter animal remains in the maternal skeleton while the other is naked at first and builds a completely new skeleton. If only nuclei and capsules divide, colonies arise. These spherical or oblong colonies, often with several constrictions, can be 4–6 cm, e.g. in the genera **Collozoum** and **Sphaerozoum** belonging to the Peripylea. Apart from binary fission, the capacity for multiple division can be seen in the Radiolaria, even in species which can also undergo binary fission (figure 43 e, f). During multiple division, numerous swarmers of equal size are formed within the central capsule; these are released as flagellated swarmers (figure 43 g) which contain one or two longish albuminoid crystals in their cytoplasm. These swarmers are reminiscent of the swarmer formation of some dinoflagellates. Their further development is not yet known. The uncertainty in this connection is aggravated by the occurrence, within or outside the radiolarian capsule, of parasitic dinoflagellates, whose various swarmers are repeatedly claimed to be microgametes and macrogametes of the Radiolaria. Such reports have led to the assumption that Radiolaria show sexuality; this has not yet been proved.

Binary fission of the **Testacea** does not begin with nuclear division but with the new growth of a second cell. There are three ways in which this can occur. When the shells are delicate and crowded closely together, the cytoplasm and shell divide longitudinally (figure 43 h, i). After this, mitosis and the distribution of the daughter nuclei begin. Forms with sturdy shells which have lost the ability to divide, e.g. *Euglypha* (figure 43 k–m), have a process of division which begins with the production of silica platelets in the cytoplasm close to the nucleus (figure 43 k). These platelets, which will serve to strengthen the shell, travel to the periphery of the protruding cytoplasm when a new cell is being formed (figure 43 l). It is only now that the nucleus prepares for division. Figure 43 m shows both daughter cells shortly before their final separation. A third possible method of shell production, which occurs, for example, in the genus *Pyxidicula,* is that in which the production of the new shell only begins when the cytoplasm has protruded to the final size of the new cell. In multinucleate Arcella species a distribution of the nuclei before the completion of their mitosis can occur; this is plasmotomy.

The agamic reproduction of the **Foraminifera** has a cyclical relationship with gamogony and should be considered together with it. The growth of the polythalamous Foraminifera is linked to the production of new chambers which are connected with one another by openings (foramina) for the passage of cytoplasm. In *Nodulina* there is only one such opening in the chamber wall at a time (figure 43 n). In contrast, figure 43 o shows, using *Elphidium crispum* as an example, numerous cytoplasmic bridges between the individual chambers; these bridges can be seen clearly by staining after the calcareous shell has been dissolved away. The production of a new chamber in *Discorbis bertheloti* is illustrated in figure 43 p and q. This foraminiferan, which is attached to a substratum, forms a mound of detritus at the periphery of its pseudopodia (figure 43 p). Within this mound, the new shell member is produced by the protruding cytoplasm, as in the Testacea. The secretion of the organic base is followed by the laying down of the calcareous deposits of the chamber wall (figure 43 q).

Sexuality in the **Amoebina** has been demonstrated with any certainly only in *Sappinia diploidea* (figure 44 a–e). This amoeba, which is normally binucleate, comes together with a second amoeba to form a common cyst coat, after which the two nuclei of each individual fuse together. This is a true sexual process (figure 44 b, c). Only after the production of the synkaryon do the cytoplasms of the now dilpoid amoebae fuse together (figure 44 d). This process is therefore a **plasmogamy**. There now follows a mitotic division linked with chromosome reduction (**meiosis**); the vegetative amoeba which is produced by this process thus has two gametic nuclei (figure 44 e, a). The binucleate nature of the amoeba is therefore due to a delay in synkaryon formation.

In the **Heliozoa** sexuality has been demonstrated only in *Actinophrys sol* (figure 33 a) and *Actinosphaerium eichhorni* (figure 32 a). In *Actinophrys sol* the axopodia first disappear (figure 44 f) and then a gelatinous coating forms. The two sister cells which arise within this coat by division of the nucleus and cytoplasm subsequently become haploid gametic cells as a result of meiosis (figure 44 g, h). The zygote (figure 44 i) is formed by the union, or **paedogamy**, of these two sister gametic cells. After a protracted resting state the diploid vegetative form is produced again from the zygote. The sexual process in *Actinosphaerium eichhorni* proceeds in basically the same way, except that the preparations here are more complex. First, most of the many nuclei disintegrate. Each remaining nucleus becomes surrounded by protoplasm and produces a uninucleate daughter animal. Paedogamy then proceeds as in *Actinophrys sol.* Finally the multinucleate condition is again established by nuclear divisions.

Sexuality in the **Radiolaria** has not yet been proved, now that it can be shown that the various swarmers are produced by dinoflagellates parasitizing the Radiolaria. Sexual processes have not yet been demonstrated in the **Testacea** either. In contrast to this, in the **Foraminifera** there is a regular cyclical **alternation of generations** between a sexual and an asexual developmental phase, **gamogony** and **agamogony**. Figure 44k–q shows this alternation using the example of *Elphidium crispum* (*Polystomella crispa*). The young uninucleate agamont (figure 44 k), which arises from the union of two gametes, becomes the central chamber of the adult **multilocular agamont** (figure 44 l) in whose cytoplasm numerous nuclei are produced by multiple agamic division; these nuclei, surrounded by protoplasm and a shell, are released as unilocular embryos (**agametes**) (figure 44 m). These agamically produced embryos grow by multiplication of chambers (n) into mature **gamonts**, i.e. mother cells of gametes. In the mature gamonts, numerous nuclei are again produced by multiple division and these, after acquiring cytoplasm and flagella, are released as gametes (figure 44 o, p). Their fusion (q) produces the zygote (k) which forms the central chamber and grows by multiplication of the chambers into the agamont, i.e. into the mother cell of the agamically produced embryos (agametes). The central chamber of the agamont which forms from the fusion of the free gametes is smaller than the shell of the agamically produced agametes which become the central chamber of the gamonts. The agamonts, as **microspheric** Foraminifera, can thereby be distinguished morphologically from the **macrospheric** gamonts (figure 44 r, s). There may also be differences in the choice of habitat, either on the sea bed or on algae in the open sea, in the gamogonous and agamogonous phases.

The life cycle of the Foraminifera can show biologically interesting variations. The course of development of *Glabratella sulcata* (figure 45 a–i) shows two important differences from that of *Elphidium crispum*. The young agamont (figure 45 a), as in *Elphidium*, changes into a multilocular adult agamont (b) but here a **nuclear dimorphism** occurs. The adult agamont contains 3 macronuclei and 9 micronuclei distributed among different chambers. When the meiosis-linked multiplication of nuclei for the production of agametes begins (c), only the generative micronuclei increase, while the 3 somatic macronuclei degenerate. If the embryos are produced within the agamont, they are released as young gamonts (e). The adult gamont is uninucleate (f). Before the nuclear division which leads to the production of gametes begins, the gamonts become paired off (g). This form of gamogony is called **gamontogamy** and differs from **gametogamy** in which the gametes are the first to seek out their sexual partners. However, the latter also occurs in gamontogamy since, after numerous gametes have been produced (h) in each gamont, the separation which at first existed between the gamonts breaks down, and the gametes seek each other out and fuse within the common outer limit of the two Foraminifera. The zygotes (i) thus produced are released as young agamonts (a), and agamogony begins again.

A further variation can be illustrated using *Rotaliella hetero-caryotica* as an example (figure 45k–t). Nuclear dimorphism is evident quite early in the young agamonts (k), three micronuclei and one macronucleus being formed. In the adult agamont (l) the somatic macronucleus lies in an outer chamber, whilst the three generative micronuclei are in the central chamber. Once again, the ensuing mitotic divisions, linked with meiosis are accompanied by the disintegration of the macronucleus (m). The agamically produced haploid uninucleate agametes (n) formed are released as young gamonts (o) and grow up to be the future uninucleate gamonts (p). Mitotic nuclear divisions (q) produce uninucleate gametes which fuse with gametes from the same gamont (r), to form uninucleate but diploid zygotes (s). This form of syngamy is called **autogamous syngamy**. Gamogony here becomes **autogamy**. Nuclear dimorphism (t) soon arises in the zygotes by nuclear divisions, so that the young agamonts (k) already possess three micronuclei and a macronucleus when they are released.

The examples used already show that, among the Rhizopoda, the Foraminifera form a group of Protozoa with special developmental characteristics, just like the *Dinoflagellata Mastigophora*, from which the Foraminifera are probably derived phylogenetically. The two possible courses of development, as seen for example in the *Coccidinia* (figure 25 a–i), with production of either agametes or of gametes by schizogony, follow one another in the Foraminifera by regular alternation. In the development of the Flagellata and of all the other Rhizopoda (see figure 241) agamogony and gamogony are two separate developmental cycles with agamogony dominant and, for many species, the sole form of reproduction, so that these classes only occasionally undergo gamogony. In contrast, there is a single combined cycle in the Foraminifera, as can be seen from figures 44 and 45. Agamogony and gamogony give rise to a cycle with **alternation of generations**. Corresponding cycles, with various modifications, characterize the development of many Sporozoa too, while the nuclear dimorphism of many Foraminifera has undergone further evolution in the Ciliata.

In the taxonomy of the Ciliata, the presence or absence of nuclear dimorphism has great significance. Our knowledge of nuclear dimorphism in the Foraminifera is, however, not yet sufficient for this to be taken into account in taxonomy. The type of gamogony, i.e. whether gametogamy or gamontogamy occurs, has similarly been given little consideration in the taxonomy of the Foraminifera, although this distinction has been important in the taxonomy of the Sporozoa.

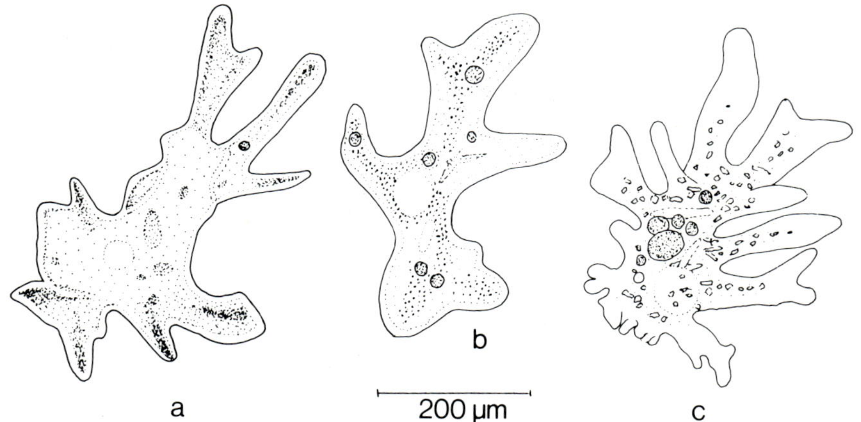

Figure 26. **Amoebina** (Magnification 105 ×)
a–c **Chaidae:**

a *Amoeba proteus* (= *Chaos diffluens*), pseudopodia in the form of lobopodia forming in all directions; on water plants and in mud
b *Metachaos discoides*, with one pseudopodium leading; in ponds and marshes
c *Polychaos dubia*, strong pseudopodia in the direction of motion. Short, sometimes tuft-like pseudopodia behind. In fresh water.

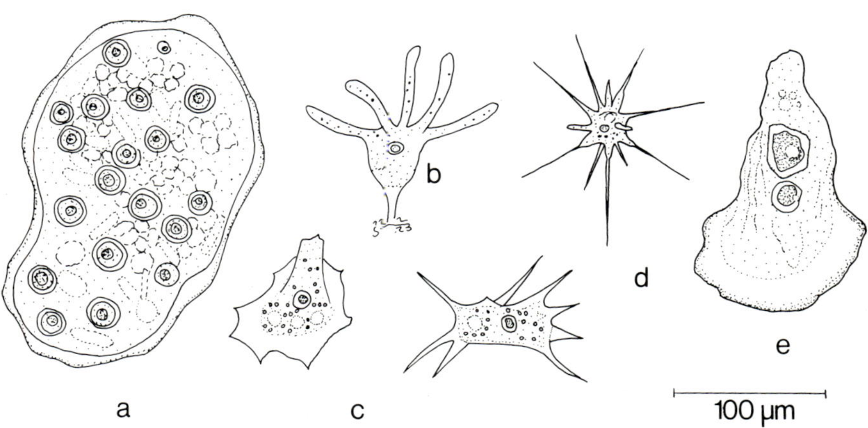

Figure 27. **Amoebina** (Magnification 205 ×)
a–b **Chaidae:**
  a *Pelomyxa palustris*, multinucleate with tough layer of ectoplasm, often with sand grains. In marshes, damp earth, moss
  b *Amoeba gorgonia*, vigorous pseudopodial formation on all sides; occasionally attached by a pseudopodium
c–d **Mayorellidae:**
  c *Mayorella vespertilio*, with filopodia, changing its shape; in mud and on water plants
  d *Astramoeba radiosa*, rigid pseudopodia; in ponds and bogs
e **Thecamoebidae:**
  e *Thecamoeba verrucosa*; in damp earth, pools, moss

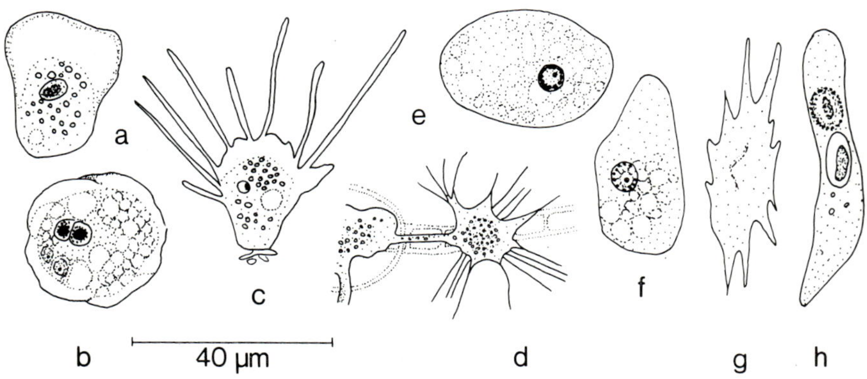

Figure 28. **Amoebina** (Magnification 690 ×)
a *Vahlkampfia guttula*, wide pseudopodium at the front; in mud and on water plants (compare *Vahlkampfia ranarum* [**Tetramitidae**] figure 16 g)
b **Thecamoebidae**:
  b *Sappinia diploidea*, with tough ectoplasm; binucleate; coprophilous
c **Mayorellidae**:
  c *Vexillifera ambulacralis*, tentaculiform pseudopodia; in small expanses of water
d **Vampyrellidae**: *Vampyrella spirogyrae*, being released from the reproduction cyst on an alga
e–f **Amoebidae**:
  e *Entamoeba coli*, intestinal parasite in man, pigmented, with phagocytized bacteria
  f *Entamoeba muris*, intestinal amoeba of the mouse
g **Acrasiae**:
  g *Dictyostelium discoideum*, a freely motile amoeba, from decaying matter
h **Paramoebidae**:
  h *Paramoeba pigmentifera*, with an additional body resembling a nucleus in front of the centrally situated nucleus; parasite of the body cavity of Chaetognatha

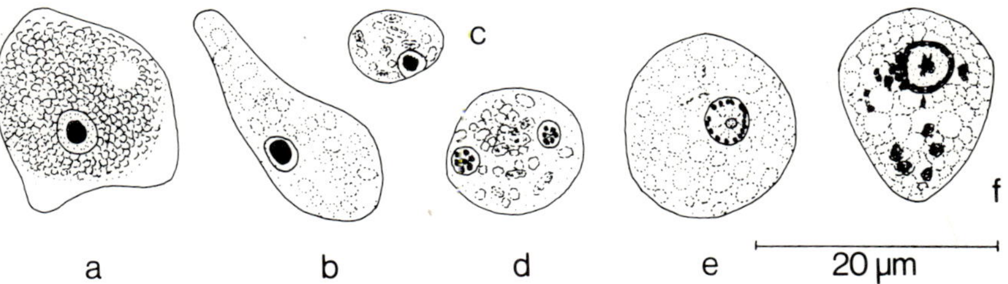

Figure 29. **Amoebina** (Magnification 1375 ×)
a–f **Amoebidae:**

a *Hartmanella hyalina* (also called *Acanthamoeba hyalina*), in waters, soil and on decaying media
b *Iodamoeba bütschlii*, intestinal parasite in man, stained nucleus compact
c *Endolimax nana*, smaller than *I. bütschlii*
d *Dientamoeba fragilis*, also an intestinal amoeba in man, here as the larger phase with 2 diffuse nuclei
e *Entamoeba histolytica*, causal agent of amoebic dysentery in man, pigmented gut-cavity form
f *Entamoeba ranarum*, intestinal parasite of frogs and toads, and their tadpoles

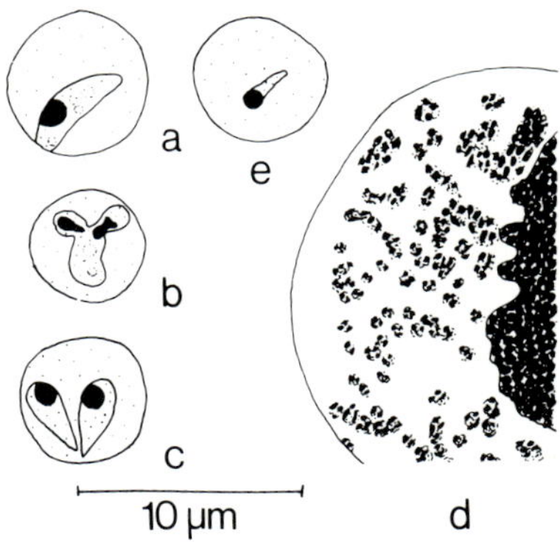

Figure 30. **Amoebina** (Magnification 2060 ×)
a–e **Piroplasmidae:**
  a–c *Babesia bigemina*, causal agent of Texas fever in cattle
  a amoeboid form
  b division
  c pear shape after division. Transmission via ticks **(Boophilus)**
  d–e *Theileria parva*, causal agent of African East Coast fever of cattle
  d reproduction in the protoplasm of a lymphocyte (only a portion shown)
  e parasite in erythrocyte; further development with transmission in the tick
  Rhipicephalus appendiculatus

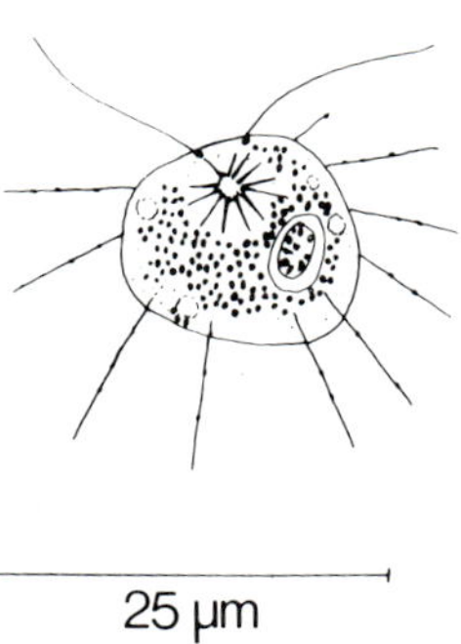

Figure 31. **Heliozoa** (Magnification 1100 ×)
  Helioflagellidae: *Dimorpha mutans*, biflagellate, with centrosome as central granule of the axopodia. In fresh water

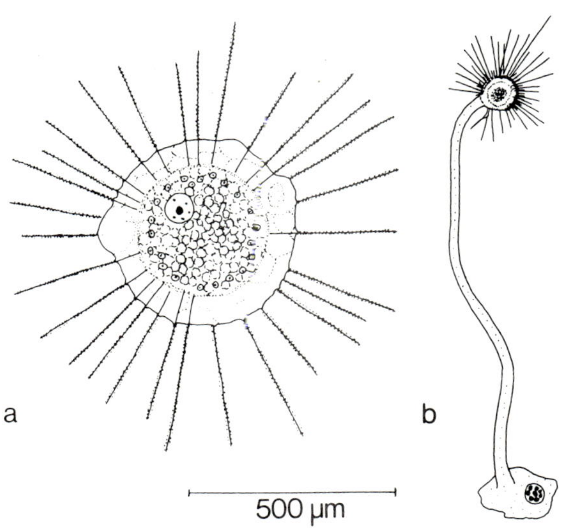

Figure 32. **Heliozoa** (Magnification 55 ×)
a **Actinophrydia:**
  a *Actinosphaerium eichhorni*, without central granule, axopodia beginning at the base of the cortex layer, central medullary substance multinucleate and with phagocytozed food. Fresh water, also stagnant water
b **Centrohelidia:**
  b *Wagnerella borealis*, attached marine form with the nucleus normally in the foot. Cell body strengthened with a gelatinous coating embedded with silica needles. Mediterranean Sea, White Sea

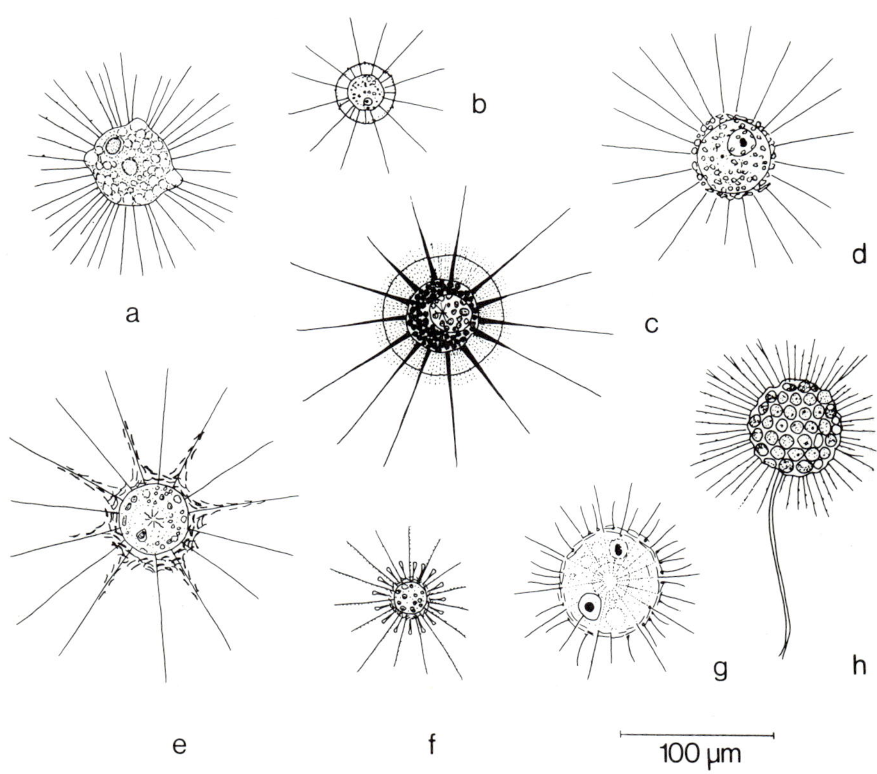
a
b
c
d
e
f
g
h
100 µm

Figure 33. **Heliozoa** (Magnification 205 ×)
á **Actinophrydia:**
   a *Actinophrys sol*, uninucleate, without a central granule, axial threads reach as far as the central nucleus. Predominantly among water plants in standing fresh water
b–g **Centrohelidia**, with central granule:
   b *Astrodisculus radians*, with gelatinous sheath without inclusions, no subdivision of the cytoplasm. Nucleus eccentric. In puddles and ponds
   c *Heterophrys myriopoda*, gelatinous sheath with numerous chitinous needles. Cytoplasm with symbiotic algae. Nucleus eccentric. In puddles, the water of marshes, and in the sea among algae
   d *Lithocolla globosa*, uninucleate, outer gelatinous layer encrusted with numerous sand grains, diatoms, etc. Reddish colour. In lakes, rivers and brackish water
   e *Raphidiophrys pallida*, outer gelatinous layer with many slightly curved scales of silica, also at the base of the axopodia. Granular ectoplasm, uninucleate. Among plants in standing fresh water
   f *Raphidocystis tubifera*, outer gelatinous layer with tangential sickle-shaped, and radial trumpet-shaped silica elements. In the water of marshes
   g *Acanthocystis aculeata*, with clear separation between endoplasm and ectoplasm. Central granule clearly visible. Gelatinous sheath with tangential silica platelets and strong radial silica needles pointing outwards. Among water plants in fresh water
h **Testacea**
   h *Clathrulina elegans*, attached, Heliozoa-like appearance, but with fine filopodia instead of axopodia, with distinction between endoplasm and ectoplasm. Central nucleus. With organic shell in the form of a perforated yellow-brownish lattice-work sphere. In small areas of water and in marshes

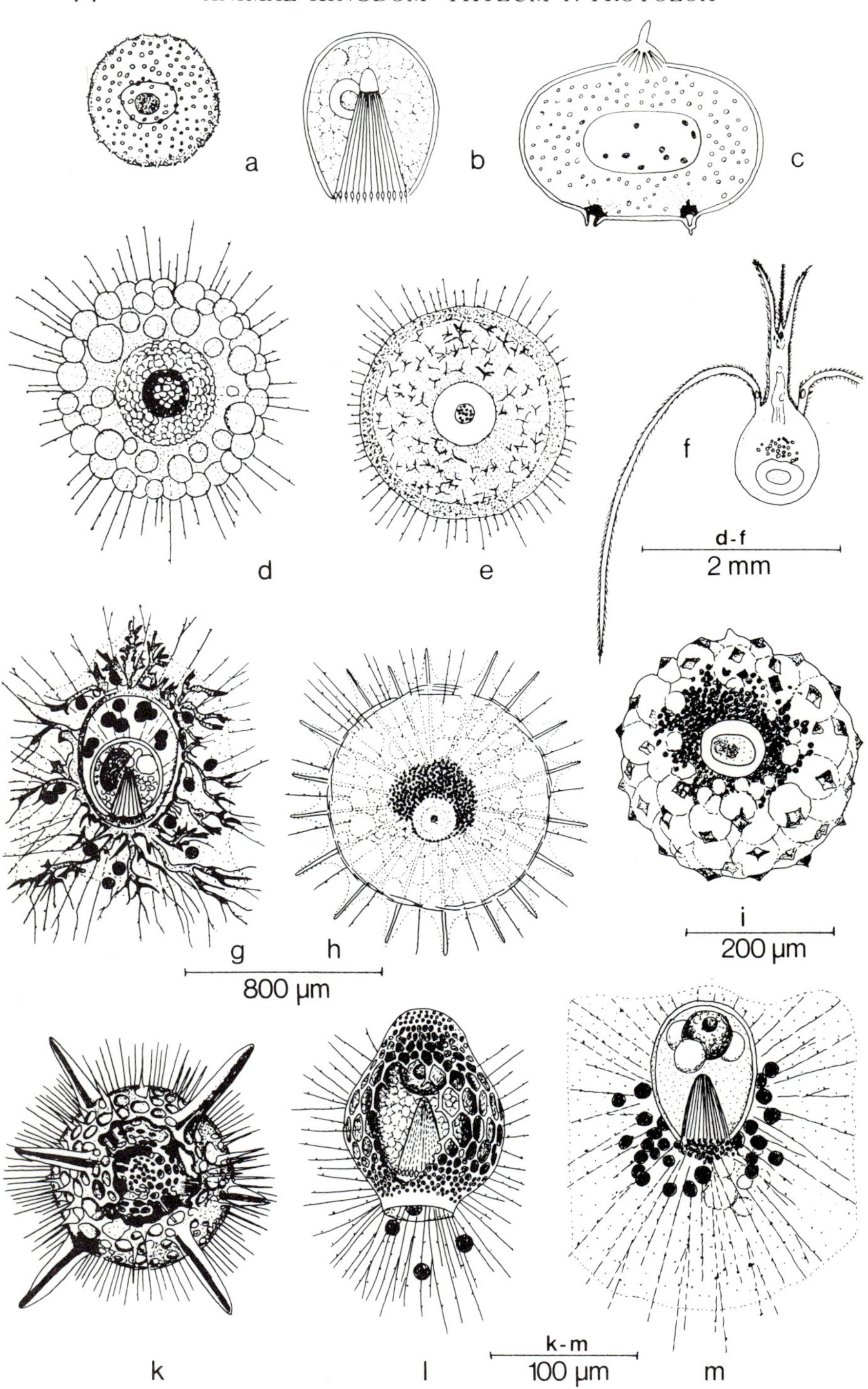
a
b
c
d
e
f
d-f
2 mm
g
h
i
200 µm
800 µm
k
l
k-m
100 µm
m

Figure 34. **Radiolaria**
a, b, c: **diagram of the central capsules**
d, e, k: **Peripylea**
g, l, m: **Monopylea**
f, h, i: **Tripylea**
a central capsule of Peripylea with pores all round and a central nucleus
b central capsule of Monopylea with pore plate, and above this the formation of a cone from the so-called "pore canals". With nucleus
c double-walled central capsule of Tripylea with astropyle and two parapyles, with large central nucleus
d–f: Magnification 14 ×
d *Thalassicolla nucleata*, with axopodia but no skeleton. The dark central capsule is surrounded by multilayered extracapsular soft bodies
e *Lampoxanthium pandora*, nucleus surrounded by vacuolated central capsule; extracapsular zone with embedded skeletal needles and gelatinous layer. With axopodia
f *Tuscarora murrayi*, central capsule with pigment in front of the astropyle. Strong silica skeleton with long (partly cut short in the figure) arm-like processes
g–h: Magnification 34·5 ×
g *Lithocircus magnificus*, central capsule with pore plate. Skeleton in the form of a rootlike ramification. With rhizopodia.
h *Aulacantha scolymantha*, skeleton composed of silica needles, some of which are hollow, radial and lying singly, while others are fine, tangential and peripheral. With axopodia. Copious phaeodium in the extracapsular cytoplasm in front of the astropyle of the central capsule.
i Magnification 105 ×
i *Caementella stapedia*, without its own skeleton, only with superficially applied foreign skeletal elements. With copious phaeodium
k–m Magnification 205 ×
k *Hexacontium asteracanthion*, skeletal from 3 latticed shells (partly removed in the figure) and radial silica needles
l *Cyrtocalpis urceolus*, central capsule with pore area. Skeleton in form of an urn-shaped lattice. With rhizopodia
m *Cystidium princeps*, without skeleton; central capsule with pore plate and nucleus as well as oil globules. Rhizopodia traverse the extracapsular gelatinous zone (only a section represented here)

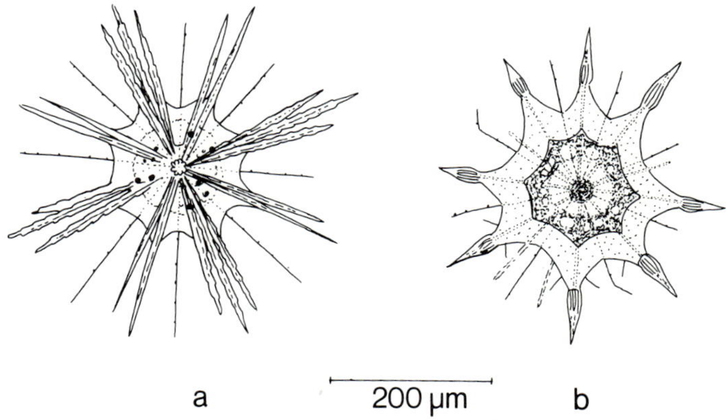

Figure 35. **Acantharia (Actipylea)** (Magnification 105 ×)
a *Conacon foliaceus*, without central capsule, with 20 skeletal needles of strontium sulphate, and with axopodia. Nuclei and (larger) zooxanthellae within the cytoplasm

b *Acanthometra pellucida*, with large multinucleate central capsule. Peripheral gelatinous zone with strong myonema established at the point where the skeletal needles emerge. With axopodia

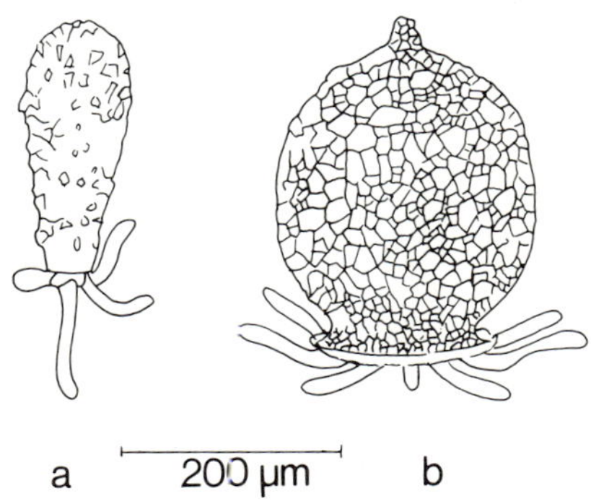

Figure 36. **Testacea** (Magnification 105 ×)
a, b **Difflugiidae:**
   a *Difflugia oblonga*, pyriform shell with xenosomes (sand grains, diatoms), terminal shell opening, slender lobopodia. In ponds, ditches, moors and damp earth
   b *Difflugia urceolata*, spherical to ovoid shell with a small peak on top, turned-up brim to the shell opening for the lobopodia. Deposits as in *D. oblonga*. In ponds, ditches, peat bogs, etc.

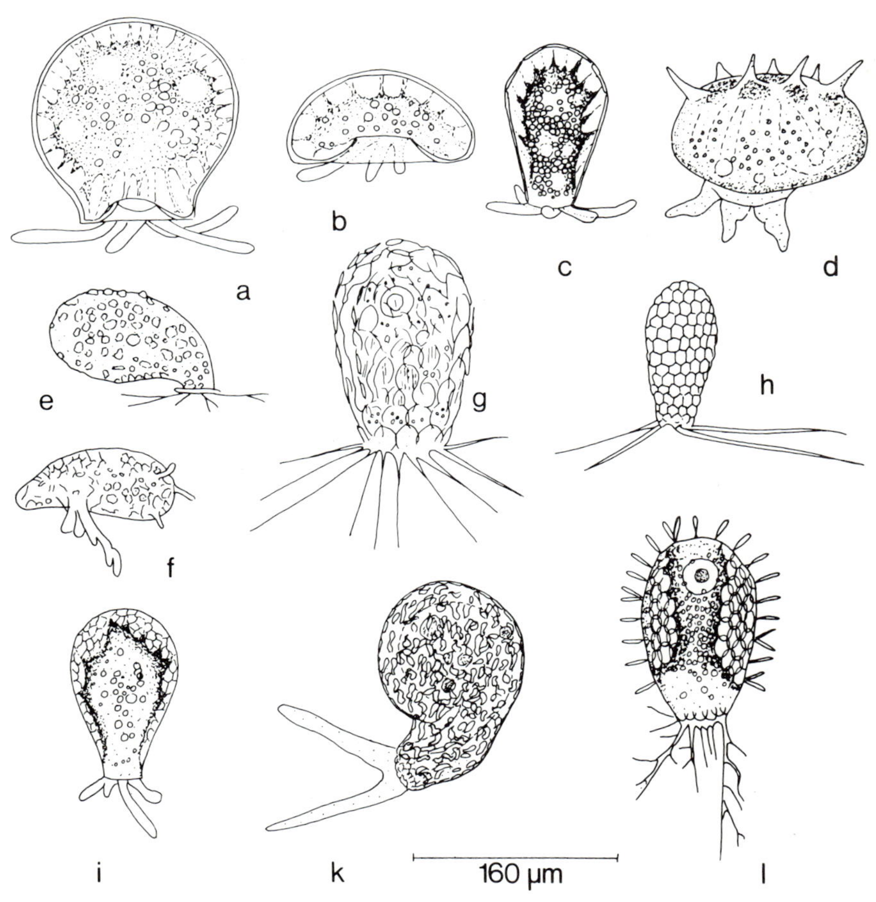

a
b
c
d
e
g
h
f
i
k
160 µm
l

Figure 37. **Testacea** (Magnification 170 ×)
a–d **Arcellidae:**
a *Arcella mitrata*, with high balloon-like organic shell with ventral funnel-shaped opening; slender lobopodia. Among water plants
b *Arcella vulgaris*, shell flatly domed with fine hexagonal markings, usually yellow to brownish. Round central opening on the convex under-side, binucleate, with several vacuoles. In standing water, mud and damp earth
c *Hyalosphenia papilio*, pyriform yellowish homogenous transparent shell with a few small pores on top. Terminal opening convex. Among plants in fresh water
d *Corycia coronata*, shell more elastic with changes in shape during movement, with 6–12 thorns. The basal opening can be drawn shut like a sack. In moss
e–f **Difflugiidae:**
e *Campascus cornutus*, bent retort-shaped shell covered with foreign bodies, predominantly quartz granules. Opening with fine transparent collar structure. Uninucleate, with filopodia. In fresh-water mud
f *Centropyxis aculeata*, cloth-cap-shaped shell with deposits of sand and diatom shells, with 4–6 thorns. Opening sunk in a depression at the base, eccentric, with finger-like lobopodia. Frequent in marsh water among algae
g–l **Euglyphidae:**
g *Euglypha aspera*, ovoid organic shell with autogenous silica platelets overlapping each other; in the region of the opening, the silica scales are finely toothed. With filopodia. In marshes, ponds, moss
h *Euglypha alveolata* (*E. acanthophora*), shell ovoid, longish, with a serrated edge projecting round the shell opening. With long filopodia. On water plants and damp mosses
i *Nebela collaris*, the thin ovoid to pyriform shell flattened at the sides, with longish oval and roundish silica platelets arranged in a mosaic. Pseudopodia finger-like, short. In peaty water and moss. The foodstuffs taken up often contain chlorophyll
k *Lesquereusia spiralis*, round, somewhat flattened shell with a neck-like region near the shell opening, and with autogenous twisted vermiform silica platelets, though partly covered with applied foreign bodies too (sand, diatoms). In the water of marshes and peat bogs
l *Placocista spinosa*, ovoid, slightly flattened, hyaline shell, covered with oval silica platelets like roof tiles, as in Euglypha, and also with silica needles. Shell opening flexible, filopodia often branched. In fresh water and bogs

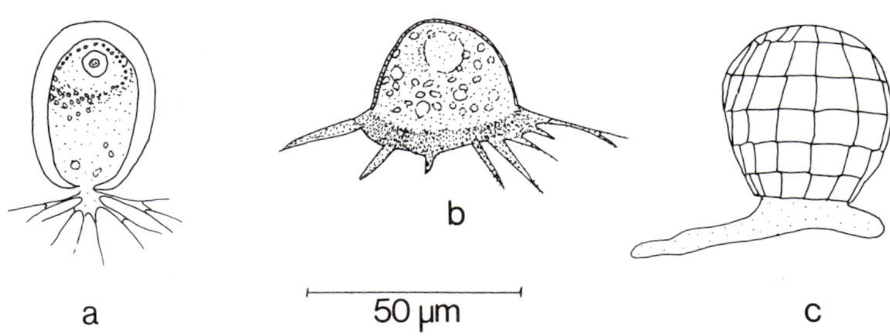

Figure 38. **Testacea** (Magnification 480 ×)
a–b **Arcellidae:**

a *Chlamydophrys stercorea*, shell thin, structureless; cytoplasm clearly segmented into zones, uninucleate, branched filopodia with pseudopodial stalk. In putrid liquids and faecal suspensions

b *Cochliopodium bilimbosum*, shell gelatinous, flexible, changing its shape. With pointed pseudopodia. In fresh water among algae

c *Quadrula discoides*, shell roundish, laterally compressed, covered in a regular pattern by square silica platelets. Shell opening with a rectilinear border, with lobopodia. In fresh water

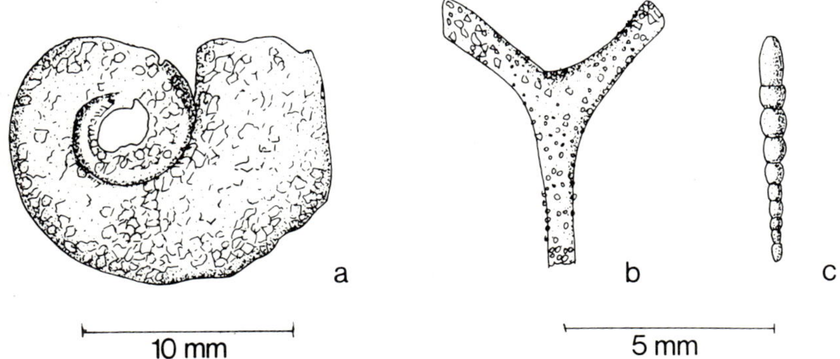

Figure 39. **Foraminifera.**
a Magnification 2·75 × . **Ammodiscidae:**
  a *Psammonyx vulcanicus*, microspheric form, terminal mouth. Shell with agglutinated sponge needles inside, sand outside, imperforate
b–c Magnification 5·5 ×
  b Rhabdamminidae: *Rhabdammina abyssorum*, tubular with an agglutinated shell, imperforate, mouth terminal
  c Nodosinellidae: *Nodulina (Rheophax) nodulosa*, shell with a linear row of chambers, coarsely sandy, imperforate, mouth terminal

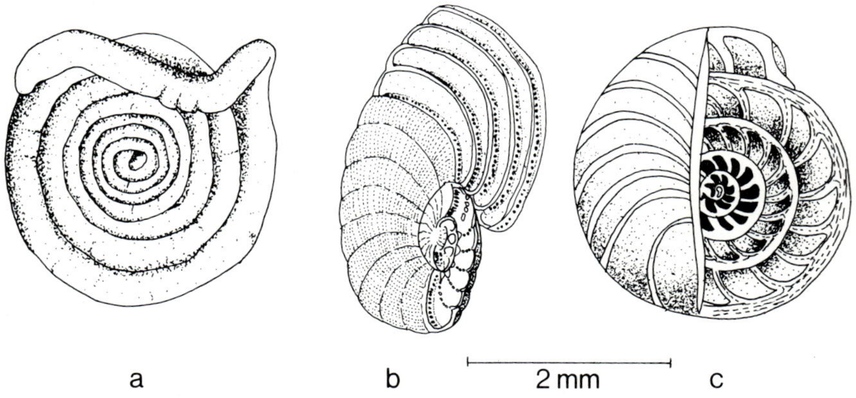

Figure 40. **Foraminifera** (Magnification 14 ×)
a **Ammodiscidae**:
  a *Ammodiscus tenuis*, monothalamous, spiral agglutinated tube with red-brown cement mass, imperforate
b **Peneroplidae**: *Peneroplis pertusus*, calcareous, polythalamous, only the first chamber perforate. Early chambers with one mouth, later chambers with several mouths
c **Rotalidae**:
  c *Nummulites cumingi*, the spirally coiled chambers exposed on the right, shell calcareous, perforate

Figure 42. **Foraminifera** (Magnification 105 ×)
a **Allogromiidae:**
a *Allogromia fluviatilis* with smooth round membranous shell, imperforate.
Cytoplasm yellowish with large nucleus and (in contrast to the marine
Foraminifera) with many contractile vacuoles. Anastomosing rhizopodia.
Mouth round (not visible here). In fresh water on plants, also in moss and
damp earth
b **Textularidae:**
b *Textularia agglutinans*, polythalamous, calcareous-sandy, alternating
chambers in two rows. Mouth cleft-shaped on the inside of the end chamber.
Perforate
c **Rotalidae:**
c *Discorbis patelliformis*, calcareous, trochospiral (compare figure 41 e),
microspheric form, with rhizopodia, perforate

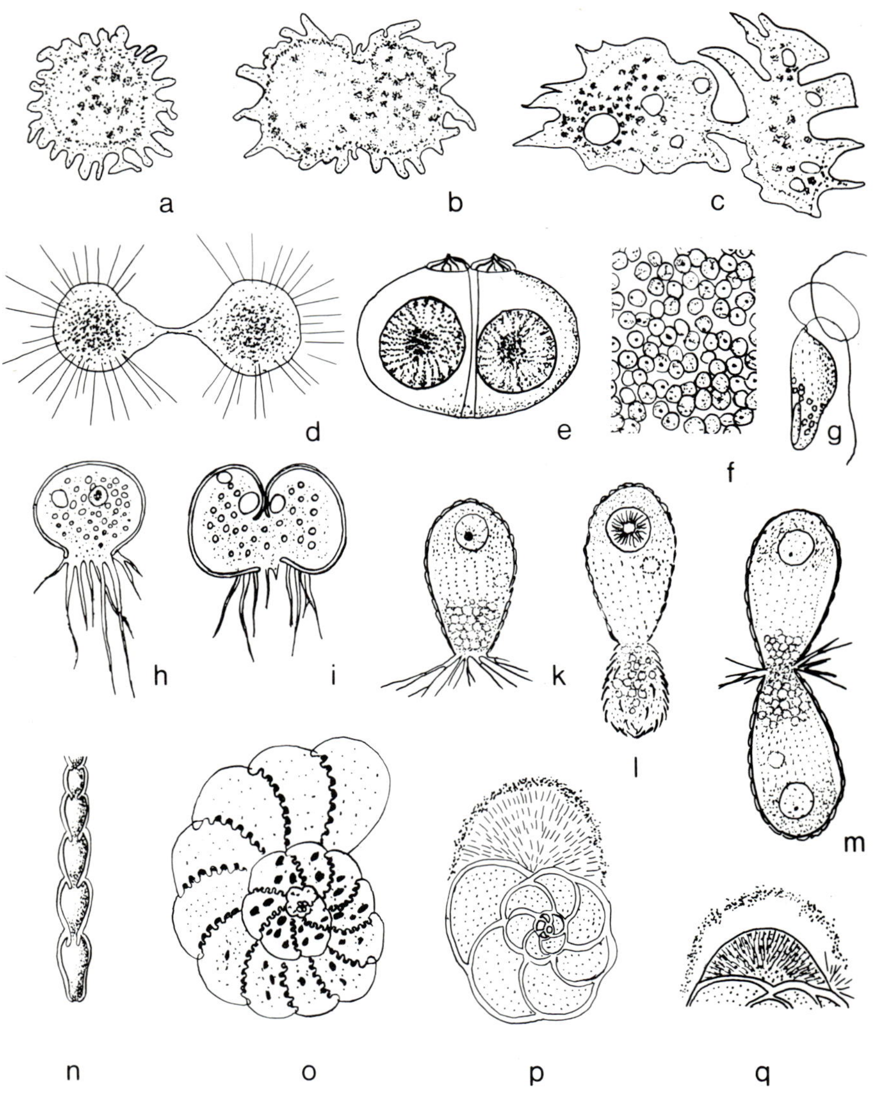
a
b
c
d
e
f
g
h
i
k
l
m
n
o
p
q

Figure 43. **Division and growth of Rhizopoda**
a–c **Amoebina**: *Mayorella vespertilio* (figure 27 c)
 a resting stage before division
 b commencement of division of cytoplasm
 c final stage of binary fission
d **Heliozoa**: *Actinosphaerium arachnoideum* (see also figure 32 a), final stage
of binary fission
e–g **Radiolaria**:
e–f *Aulacantha scolymantha* (figure 34 h)
 e binary fission of the central capsule
 f part of the inside of a central capsule with formation of crystal swarmers
 g *Collozoum fulvum*, crystal swarmer
h–m **Testacea**:
h–i *Cochliopodium pellucidum* (see also figure 38 b)
 h resting condition
 i binary fission with division of the shell
k–m *Euglypha alveolata* (figure 37 h)
 k resting stage with reserve platelets for shell production in the cytoplasm
 near the nucleus
 l formation of a cytoplasmic bud with reserve platelets laid down peripherally
 m two daughter animals with completed shell formation shortly before the
 conclusion of binary fission
n–q **Foraminifera**:
n–o showing foramina
 n *Nodulina (Reophax) nodulosa* (figure 39 c) with only one shell-opening at
 a time between the chambers
 o decalcified and stained soft part of the microspheric form of *Elphidium
 crispum (Polystomella crispa)* showing the numerous cytoplasmic bridges
 between the individual chambers (see also figure 44 r, s)
p–q *Discorbis bertheloti*, formation of a new chamber
 p production of a mound of detritus through the rhizpodia
 q production of the organic foundation of the new chamber wall and subsequent calcareous deposition within the mound formation.

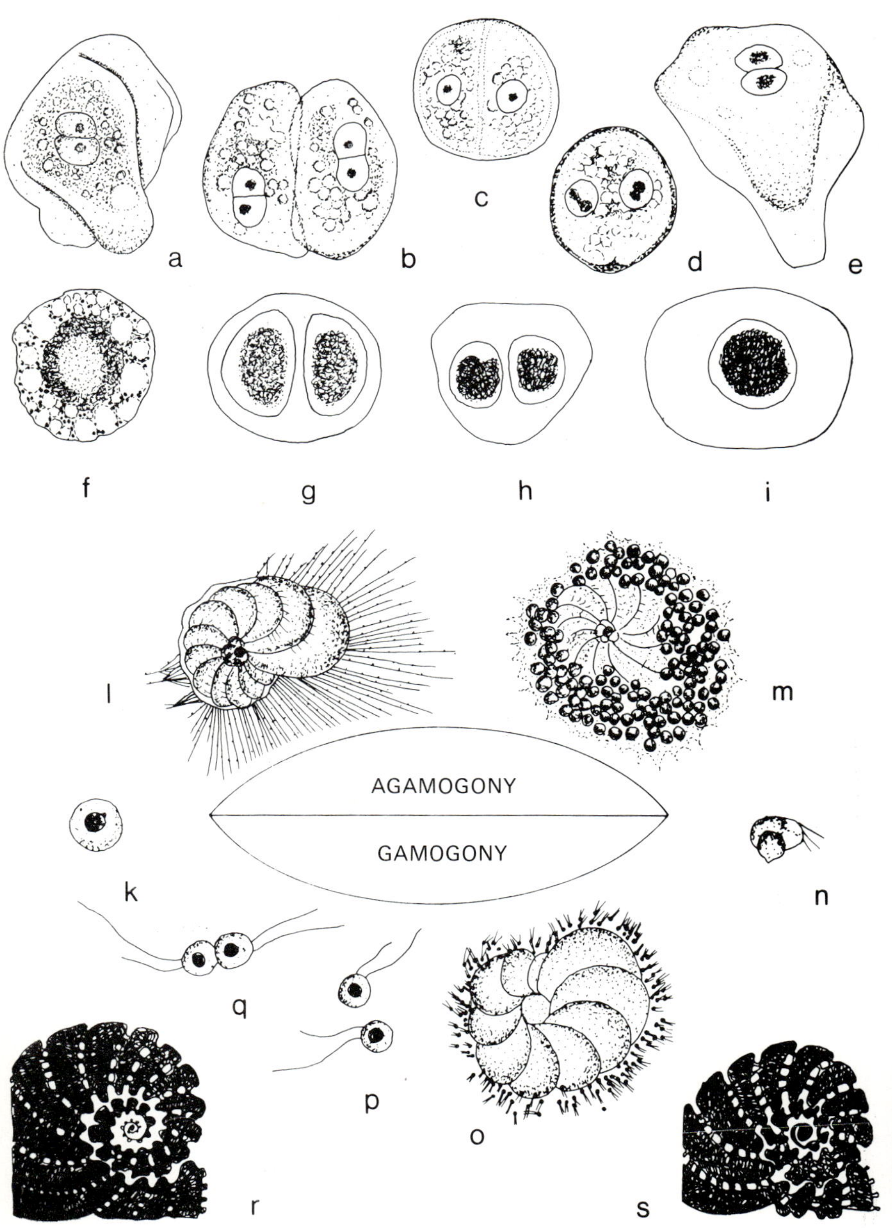
a
b
c
d
e
f
g
h
i
l
m
AGAMOGONY
GAMOGONY
k
n
q
p
o
r
s

Figure 44. **Sexuality in Rhizopoda**
a–e **Amoebina**: *Sappinia diploidea* (figure 28 b)
  a normal vegetative stage with 2 gametic nuclei
  b encystment of two amoebae
  c fusion of the gametic nuclei in each cell
  d fusion of cytoplasm of the amoebae (plasmogamy) with the onset of reduction division of the nuclei (meiosis)
  e re-released amoebal stage with 2 gametic nuclei
f–i **Heliozoa**: *Actinophrys sol* (figure 33 a)
  f retraction of the axopodia
  g division into two daughter animals
  h the developed gametes
  i zygote (drawn from a living specimen)
k–s **Foraminifera**: developmental cycle of *Elphidium crispum* (*Polystomella crispa*)
  k young agamont
  l adult agamont with rhizopodia
  m emerging embryos (agametes) = young gamonts
  n onset of growth of the gamonts
  o release of the flagellate gametes
  p free gametes (higher magnification)
  q fusion of the gametes (isogamy)
  k zygote = young agamont
r–s *Peneroplis pertusus* (figure 40 b), central portion of the decalcified foraminiferan
  r agamont = **microspheric** form
  s gamont = **macrospheric** form

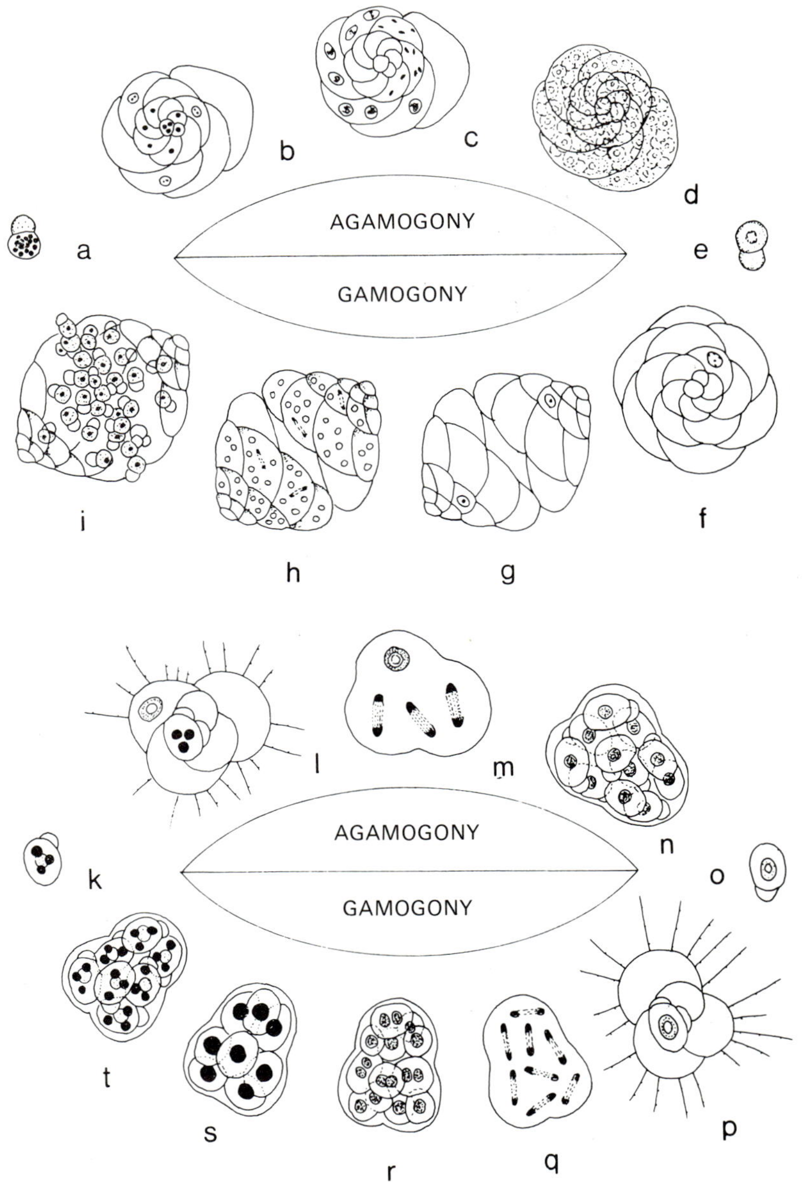
AGAMOGONY
GAMOGONY
a
b
c
d
e
f
g
h
i
AGAMOGONY
GAMOGONY
k
l
m
n
o
p
q
r
s
t

Figure 45. **Sexuality in Rhizopoda**: development of Foraminifera
a–i *Glabratella sulcata*
  a young agamont
  b adult agamont with 9 micronuclei and 3. macronuclei
  c onset of division of the micronuclei (meiosis)
  d agamont with numerous agametes
  e free agamete = young agamont
  f adult gamont
  g two gamonts lying against one another (syzygy formation)
  h onset of production of the gametic nuclei
  i liberation of the zygote
  a free zygote = young agamont
k–t *Rotaliella heterocaryotica*
  k young agamont with 3 micronuclei and 1 macronucleus
  l adult agamont (3 micronuclei, 1 macronucleus)
  m onset of division of the micronuclei (meiosis)
  n agamont with agametes
  o free agamete = young gamont
  p adult gamont
  q gamont with onset of production of the gametic nuclei
  r fusion of the gametes (autogamy)
  s gamont with zygotes
  t gamont with young agamonts, each with 3 micronuclei and 1 macronucleus
  k free young agamont.

CLASS III: SPOROZOA (LEUCKART 1879)
SPORE ANIMALS

ALL SPOROZOA ARE PARASITES WHICH PRODUCE SPORES. THE spores, often enclosed in a strong coat, transmit the infection. Their production varies according to subclass; in the Telosporidia a number of spores is usually produced within an oocyst. However, there are also genera which produce with only a single spore, e.g. the coccidian genera *Caryospora, Lankesterella* and *Mantonella*. In the absence of an additional spore coat, the oocyst itself can become the spore, as in *Haemogregarina stepanowi* (figure 55), *Pfeifferinella, Schellackia* and *Cryptosporidium*. This is also true for the *Haemosporidia* which are transmitted by alternative hosts (figures 59, 60).

SUBCLASS 1: TELOSPORIDIA (Schaudinn 1900)

In the Telosporidia spore production comes at the end of the development of the individual. Like the Foraminifera, the Telosporidia have an alteration of generations with agamogony and gamogony in a cyclic sequence in which the zygote resulting from the fusion of two gametes leads to spore production. The embryos which form agamically in the spore are the sporozoites, and the development of the spores is known as **sporogony**. The sporozoites may grow first into multinucleate schizonts which dissociate into uninucleate merozoites to disseminate the infection in the affected host or, alternatively, they can develop directly into gamonts. Further differences in the development of the gamonts leads to the taxonomic separation of the Orders **Gregarinida** and **Coccidia**.

Order 1: Gregarinida (Lankester 1866), figures 46–52

The Gregarinida are parasites of the gut or body cavity of invertebrates, and are found predominantly in annelids and arthropods. They are extracellular but, in the early stage, can develop sometimes temporarily intracellularly. In the adult stage they can often reach a considerable size with a markedly individual appearance (figures 47, 49, 50, 52). The name they have been given (Gregarinida = gregarious animals) reflects the occurrence in the infected host of accumulations of their cells as they assemble in gamontogamy. The **Eugregarinida** show agamogony only in the form of sporogony, without subsequent schizogony; in this respect they differ from the **Schizo-gregarinida**.

Suborder 1 : Eugregarinida

In the Eugregarinida, the "true" Gregarinida, the sporozoites develop directly into the large gamonts without additional schizogony. The cell body may or may not be segmented; this criterion characterizes the families **Polycystidae** (=**Cephalina**) and **Monocystidae** (=**Acephalina**) respectively. An example of the developmental cycle of the Eugregarinida is illustrated in figure 46 by the monocystid *Lankesteria culicis* from the gut of the mosquito *Aedes aegypti*. Figure 46 i shows the penetration of the sporozoite into the intestinal epithelium of a mosquito larva and the growth of the gamont, which in the adult stage lies free in the gut lumen (k). The gamogony which then follows takes the course of gamontogamy, in which two gamonts at a time associate together in syzygy, and then encyst together within the Malpighian tubules when the mosquito larva pupates (l). Schizogony leads to the formation in each gamont of numerous peripheral gametic nuclei (m), from which the gametes form when the wall of the gamont disintegrates (n). Numerous zygotes, the young spores (a), are produced by the fusion of pairs of gametes within the common cyst. In the course of the subsequent sporogony, the initially uninucleate spores become 8-nucleate through agamic nuclear multiplication (b–f). Each of the eight nuclei produces a sporozoite (g, h), which again infects the gut epithelium (i) when the spore is ingested by a mosquito larva.

The **Monocystidae** are predominantly parasites of the body cavity and seminal vesicles of oligochaetes, e.g. in the seminal vesicles of earthworms (figure 47 a, d, e). Here they develop first in the blastophores, but are later released by the disintegration of the host cells. *Enterocystis ensis* (figure 47 b), on the other hand, lives in the gut of larvae of mayflies of the genus *Caenis*, while *Urospora chiridotae* (figure 47 c) lives in the blood vessels of the holothurian *Chiridota laevis*. In some species (see figure 47 c–e) the surface of the cell body is entirely or partially covered with cell-derived filiform appendages. The gamonts sometimes undergo syzygy (figure 47 b, c) prematurely, before their common encystment.

The developmental cycle of the **Polycystidae**, which are almost exclusively parasites of arthropods, corresponds basically to the development of the Monocystidae, but the gamont growing up from the sporozoite is at first tripartite, as can be seen, for example, in *Gregarina cuneata* from the gut of the mealworm (figure 48 a–d). The parts are differentiated antero-posteriorally, as epimerite, protomerite and deutomerite. The nucleus always lies in the deutomerite. The epimerite serves to anchor the growing gamont and is therefore an attachment organelle; it can have very diverse forms in the different species of Gregarinida (figure 48 e–m). In a later phase the epimerite is discarded, so that the organisms are now only bipartite, with the protomerite and deutomerite (figure 49 a–d). Syzygy in relation to gamontogamy can occur prematurely, before the common encystment, in the Polycystidae too (figures 48 n, 49 e), so that both syzygites remain motile together; in this case the anterior partner is called the **primite** and the posterior partner the **satellite**.

Gregarina ovata (figure 48 n–q) from the gut of the earwig *Forficula auricularia* is an example of the further development of the life cycle showing gamogony, encystation (o) and production of gametes which then lie in a peripheral position for a time (p) and which, when they unite with one another, produce the zygotes, i.e. the young spores (q), within a common cyst. The cyst containing the spores is called the **sporangium**. The spores escape through tubular structures, the **sporoducts**, which protrude when ripe. The structure of the sporangia with their sporoducts is reminiscent of the structure of some dinoflagellates (figure 25 c, g). Material easily obtained for the study of Polycystidae is larvae of the mealworm beetle *Tenebrio molitor*. Four species can be found in the gut of the mealworm (figure 49 a–e): *Gregarina polymorpha* (a), *G. cuneata* (b, e), *G. steini* (c) and *Steinina ovalis* (d). Differences in the form of the Polycystidae are shown in figures 49 f and 50 a–e, in which epimerites are also sometimes shown.

Suborder 2 : Schizogregarinida

The agamic reproduction of the Schizogregarinida is not limited to the formation of sporozoites before infection as it is in the Eugregarinida; reproduction of the parasite in the infected host is achieved by schizogony with merozoite formation. The difference in development when compared with the Eugregarinida cycle (figure 46) is clear from figure 51. The example used, *Schizocystis gregarinoides* from the gut of the dipteran larva *Ceratopogon solstitialis*, shows that, as in the Eugregarinida, the sporozoites (b) develop from zygotes lying in a cyst (figure 51 a) and become attached to the gut wall by their anterior ends (c). The schizonts can grow to 400 μm and contain up to 200 nuclei (d) as a result of nuclear multiplication. The uninucleate merozoites (e) produced from these can again develop into schizonts on the gut wall (f, d) or they become gamonts (e, g), which in the gamontogamic sense undergo syzygy (h) and encyst together (i). The gametes which form around the periphery of the two gamonts are morphologically different (k). The subsequent fusion is therefore anisogamous fertilization giving rise to zygotes within a communal cyst (a).

The Schizogregarinida are parasites of the gut and organs connected to this in arthropods, annelids and tunicates. The parasites attach themselves to the gut wall by means of an anterior zone of cytoplasm called the **pseudomerite** which is not, however, separated by a dividing wall as an epimerite would be. The shape of the parasite before the onset of schizogony is the same as that of the adult form. Figure 52 provides some examples of the variability in size and shape found in the Schizogregarinida.

Order 2: Coccidia (Leuckart 1879), figures 53–64

The Coccidia are "berry-shaped", round to oval Sporozoa, as the name implies. They differ from the Gregarinida in the behaviour of the female gamont which does not give rise to a number of gametes, but itself constitutes a large gamete without dividing. In contrast, the development of the male gamont proceeds as it does in the Gregarinida: a number of smaller gametes are produced from it as a result of nuclear divisions. Syngamy thus becomes oogamy, in which a nutrient-rich largely immotile macrogamete is fertilized by a small flagellated motile micro-gamete. The large zygote which is produced by oogamy is known as the **oocyst**. The spore nuclei, and hence the spores, are formed in the oocyst by a first division phase of the synkaryon. The subsequent production of sporozoites within each spore is like that occurring in the Gregarinida, and these either grow directly into gamonts or else give rise to multiplication of the parasite within the infected host by means of schizogony. This has led to the creation of two suborders, **Protococcidia** and **Schizococcidia**. In the suborder **Toxoplasmida**, a further development in the form of agamic reproduction seems to occur.

Suborder 1 : Protococcidia

The Protococcidia are Coccidia with a predominantly extra-cellular mode of life in the body cavity of marine worms (Archiannelida, Polychaeta). Agamogony proceeds, as in the Eugregarinida, as sporogony alone without subsequent schiz-ogony. Figure 53 illustrates this cycle by reference to *Eucoccidium dinophili* from the body cavity of the transparent archiannelid *Dinophilus gyrociliatus*. The gamonts (f, h) develop directly from the sporozoites (e). Multiple division in the micro-gamont (h) gives rise to 12–32 biflagellate microgametes (i–l), while the macrogamont (f) becomes the macrogamete (g) without further divison and is fertilized by a microgamete (m). Meiosis occurs with the elongation of the synkaryon in the zygote (a). Agamic multiple division within the oocyst leads to production of spore nuclei (b) and then the spores (c), which are uninucleate at first and in each of which 6 sporozoites are produced by further divisions (d, e).

Suborder 2 : Schizococcidia

In contrast to the Gregarinida, most species of the Coccidia have an additional cycle of schizogony. The Schizococcidia live almost exclusively intracellularly. The number of agamic schizogonous generations is constant in many species and is determined genetically, not by the influence of the host. Differences in the development of the Schizococcidia lead to the differentiation of the three superfamilies **Adeleidea**, **Eimeridea** and **Haemosporidia**.

Gamogony proceeds as gamontogamy in the **Adeleidea**, just as in syzygy in the Gregarinida. The Gregarinida always have a single host, but the Adeleidea can have an alternation of hosts. Figure 54 shows an example of a cycle without an alternation of hosts, *Adelea ovata* from the gut of the centipede *Lithobius forficatus*. The sporozoite (figure 54 d) penetrates into the gut epithelium and develops into a rounded uninucleate coccid (e) in which schizogony produces the merozoites, each being about 20 µm (f, g). These can infect fresh epithelial cells (h) and repeat the schizogony, or else they become the gamonts (i) which, in the Schizococcidia, are also called **gametocytes**. The microgamont positions alongside a macrogamont and produces four microgametes, one of which fertilizes the macrogamete (k–m); this then becomes the oocyst (a). The remaining microgametes degenerate outside the membrane of the oocyst. Multiple division during sporogony within the oocyst gives rise to as many as 32 or more spores (a–c), each one subsequently producing 2 sporozoites (c).

Several genera have a cyclical **alternation of hosts** which complicates the developmental cycle, with the alternation usually between vertebrates and arthropods (**Karyolysus** in lizards, **Hepatozoon** in rodents, etc.) or, in the case of *Haemogregarina stepanowi* (figure 55), with the European marsh-turtle *Emys orbicularis* as vertebrate host and the proboscid leech *Placobdella catenigera* as invertebrate carrier. The sporozoite (figure 55 d) reaches the blood of the turtle during sucking by the leech. Schizogony begins in the red corpuscles of the bone marrow with the formation of usually 24 merozoites (e, f), which infect more erythrocytes and repeat the schizogony. This time the schizonts remain smaller, with only 4–6 merozoites, and are also found in the peripheral blood (g–i). The merozoites which occur in the blood can also undergo schizogony (k), or else they become schizonts whose merozoites are significantly smaller (l, m) and become young gamonts (= gametocytes) (n, p), which also grow in the erythrocytes; the macrogamont is rich in foodstuffs (o), while the microgamont has an enlarged nucleus and anterior striation (q). After the leech has taken up the infected blood, the blood corpuscles disintegrate into the gut and the gamonts undergo syzygy (r), after which the microgamont produces 4 microgametes (s). One of these fertilizes the macrogamete (t), and nuclear division in the zygote (oocyst) gives rise to 8 sporozoites (a–c). An additional spore coat is acquired during sporogony in the oocyst. The sporozoites are transmitted to another turtle during the process of sucking. Figure 55 shows clearly that the stage of the life cycle of the parasite in the [peripheral] blood is prolonged as a result of increased agamic schizogonous activity (e–p) and that gamogony in the invertebrate host shows no variations in its form.

The **Eimeridea** have no cycle of gamontogamy as occurs in the Gregarinidae and Adeleidea; instead they undergo gametogamy. The macrogamonts and microgamonts (= gametocytes) develop separately, and the microgametes have to locate the macrogametes for fertilization to occur. For this reason, the microgametes produced are much more numerous than the macrogametes. The Eimeridea are widely distributed in numerous invertebrate and vertebrate animals, both on land and in water, and occur predominantly as parasites of the gut and of organs connected with it. Many host animals can be infected by several species of Coccidia, but the parasites themselves have a narrow host specificity.

The Eimeridea also include species with and without an alternation of hosts. Figure 57 shows an example of a cycle without an alternation of hosts, *Eimeria schubergi* from the gut of the centipede *Lithobius forficatus*. The sporozoite (d) penetrates into the gut epithelium and becomes an agamont (e). Schizogony gives rise to merozoites (f, g) which at first infect other epithelial cells. After repeated schizogony, young gamonts (i, m) form the merozoites. The macrogamont (i) grows directly into a uninucleate macrogamete (k, l), while the microgamont (m) produces flagellated microgametes (n–p). The latter swarm round the macrogamete, although only one can penetrate and undergo karyogamy. The zygote formed as a result of fertilization produces an outer coating, and in this state is known as an oocyst (a). Two nuclear divisions give rise to 4 spores (b) in the oocyst, each of which produces 2 sporozoites (c).

Distinction between species is not based upon the morphology of the schizonts (figure 57 e), but upon the morphology of the oocyst and its spores which are eliminated with the faeces; the species of infected host animal is also taken into account. Figure 56 shows some examples of Eimeridea of the genus **Eimeria** with 4 spores in the oocyst, each with 2 sporozoites, and of the genus **Isospora** with 2 spores in the ripe oocyst, each spore with 4 sporozoites. A member of this genus, *Isospora belli*, is a parasite of man (figure 56 r, s). One member of the Eimeridea with **alternation of hosts** is *Aggregata eberthi* (figure 58). The sporozoites become schizonts (e–h) in the intestine of the crab *Portunus depurator*. The merozoites (h, i) penetrate the gut wall of the cuttlefish *Sepia officinalis* when this eats an infected crab, and there they become gamonts. The microgamont produces biflagellate microgametes (m–o) by schizogony and these fertilize (p) the large macrogametes. Meiosis takes place in the zygote (a) and spores (b, c) are produced by agamic nuclear division. Three sporozoites form in each spore, and another crab is infected by ingesting the spores.

The **Haemosporidia** are Eimeridea which show an alternation of hosts and a secondary cycle of parasitism in the blood; the primary infection of the vertebrate host remains an infection of tissues, like that occurring in the other Eimeridea. The vectors are blood-sucking insects in which the zygote of the parasite first develops into a motile vermiform ookinete.

In the family **Haemoproteidae**, in which many of the species are parasites of birds, only the young gamonts infect the blood corpuscles, while schizogony takes place only in the tissues, as shown in figure 59, which illustrates infection in the domestic pigeon *Columba livia* by *Haemoproteus columbae*. The sporozoites, which are transmitted by the louse-fly *Lynchia maura*, penetrate into the endothelial cells of the pigeon's vascular system (c) and at first multiply to produce several uninucleate parasites (d, e) which then grow into schizonts (f). All of these schizonts form numerous merozoites (g) which infect the nucleated red blood corpuscles and there become macrogamonts (h, i) or microgamonts (l, m). Further development occurs at first in the gut lumen of a louse-fly which has taken up some of the infected blood. The macrogamont becomes a macrogamete (k); microgametes (n) are released from the microgamont and fertilize the macrogamete (o). The zygote becomes a motile uninucleate ookinete (p) and penetrates into the gut wall. In the peritoneum it changes into an oocyst (a) in which numerous sporozoites are produced without the formation of additional spore coats (b). After rupture of the ripe oocyst, the sporozoites reach the salivary gland of the louse-fly by way of the haemolymph. In contrast to the genus **Haemoproteus**, which occurs in birds and reptiles, the gamonts of the genus **Leucocytozoon** infect the white, not the red, blood corpuscles in birds. The genus **Hepatozoon** occurs in some mammals, including apes.

The family **Plasmodiidae** includes the malaria parasites, where schizogony takes place in the blood of the vertebrate host, in a similar way to the development of the Adeleidea (figure 55). This blood parasitaemia is illustrated by reference to tertian malaria infection in man caused by *Plasmodium vivax* (figure 60). Mosquitoes of the genus *Anopheles* act as vectors; when they bite, the sporozoites (e) pass to the human liver within an hour via the blood. Schizogony begins in the parenchyma cells of the liver (f, g), and the merozoites produced can infect more liver cells (h); this exoerythrocytic schizogony can last for an average of $1\frac{1}{2}$–2 years in *P. vivax* infections. Some 8 days after the infection has taken place, merozoites from the liver (i) also penetrate the peripheral red blood corpuscles. This initiates the cycle of erythrocytic schizogony, which in *P. vivax* generally leads to production of 16 merozoites (j–l), while an exoerythrocytic schizont produces many thousands of merozoites. The maturation of the schizonts in the blood takes 48 hours in *P. vivax*, and the merozoites released by disintegration of the mature blood schizonts can lead to a repetition of schizogony in the blood (m), or they tend increasingly to form macrogamonts (n–p) and microgamonts (r, s), whose further development will only take place in the stomach of the *Anopheles* mosquito. Linguistic usage in a clinical context has led to the gamonts in malaria being known as gametocytes. In the mosquito, microgametes (t) form from the microgamont (microgametocyte) which has a diffuse nucleus. The microgametes unite (u) with the macrogamete, and again a zygote, in the form of a motile ookinete, is produced (v) and becomes an oocyst on the mid gut of the mosquito (a). The sporozoite nuclei (b) are formed here by agamic division without additional formation of spores and, finally, several thousand sporozoites (c), each measuring about 15 μm, are produced and reach the salivary gland of the mosquito by way of the haemolymph (d). The development of the malaria parasites is often linked epidemiologically to particular climatic zones or seasons of the year, because of the temperature required for development within the mosquito. This is true for four species of malaria which occur in man (figure 61 a–m). In addition to the infection of man and apes by various members of the Plasmodiidae, malaria infections occur in rodents, in reptiles and, in particular, in birds. Species differentiation is based upon the morphology of the erythrocytic phase in the vertebrate host, the duration of each cycle of this phase, and the species of animal able to be infected. Figure 61 demonstrates differences in the morphology of the erythrocytic forms of the parasites which affect man and of some parasites of birds.

Suborder 3 : Toxoplasmida

Two groups of parasites of vertebrates will be described here as examples of the Toxoplasmida, namely *Toxoplasma* and the *Sarcosporidia*. In their primary hosts, both have a cycle of development like that of the Eimeridea, but multiply in other hosts only by a special form of agamogony, without gamogony.

Toxoplasma gondii is the only species of the genus **Toxoplasma** to have been recognized so far. The coccidial life cycle proceeds in the gut of cats of the genus *Felis*, particularly in the domestic cat *F. domestica*, which can thus be regarded as the major host. The sporozoite (figure 62 e) initiates schizogony of the parasite in the intestinal wall of the cat (f–h). The first-cycle merozoites (i) may initiate a repetition of the schizogony cycle (i) or they may become gamonts (j, k, n). The macrogamont (k) develops into the macrogamete (l, m), and the microgamont (n) undergoes schizogony to produce microgametes (o, p). The biflagellate microgamete (q) fertilizes the macrogamete (m), and the zygote so formed becomes an oocyst with 2 spores (b). Four sporozoites are produced in each spore (c, d). The Toxoplasmida thus have a cycle characteristic of the genus *Isospora*, but differ from the Eimeridea by their capacity for further agamic development in numerous other vertebrates—most frequently warm-blooded animals but, when the temperature is adequate, in vertebrates with a variable body temperature. In these secondary hosts the sporozoite (e) develops into a sickle-shaped or "bow-shaped" trophozoite (r) 7–8 µm long; it is from this form that the name of the parasite is derived. In numerous organs, in particular in the cerebral region of the brain, endogenous binary fission (s) leads to agamic reproduction, or else large cysts are formed (t) in which numerous smaller toxoplasms are produced. The infection can spread among the secondary hosts, without the intervention of the coccidial phase of development in the cat, through transmission via the placenta, and orally via infected meat, mother's milk and excrement; for this reason infected domestic animals can be particularly dangerous for man. The primary host, the cat, is infected in this way predominantly via rodents carrying toxoplasms.

The great lack of specificity, with regard to both organs and host, shown by the parasite in the secondary host is most unusual in the Coccidia. Perhaps the special endogenous type of division of the parasite assists in this, by enabling the young stages to develop through the protection afforded by the mother cells. This type of binary fission is known as **endodyogeny**. Figure 63 illustrates this process, first through the division of the nucleus (a–c) and then the intracellular production of the daughter cells (d, e). The free daughter cells (f) correspond to figure 62 s. In the schizonts in the gut of the cat, development from the 4-nucleate stage proceeds according to a pattern of division characteristic of endogeny, each individual probably giving rise to 32 daughter forms. This type of division is known as **endopolygeny**.

Recent studies have shown both endogeny and the presence of a coccidial cycle in the **Sarcosporidia**. The Sarcosporidia are parasitic in the muscles of various mammals, particularly the larger domestic animals, and of birds and reptiles. Man too can occasionally be infected. In the striated muscular system, intracellular cysts are formed which can become several centimetres long in some species. However, the Sarcosporidia, like *Toxoplasma*, are primarily intestinal Coccidia, and in this case even the coccidial cycle is not always confined to a single host genus. Thus *Sarcocystis fusiformis* from the muscular system of cattle can develop in the gut of dog, cat and man. Figure 64 a and b show a thin-walled oocyst from a dog and a man respectively; each oocyst contains 2 spores. In the same way *Sarcocystis miescheriana* from the muscular system of the pig is able to develop as a coccidian in the human intestine (figure 64 d), while the cat has been shown to be the primary host of *S. tenella* (c) from the sheep. The oocysts obtained in this way from the human intestine (b, d) with 2 spores each containing 4 sporozoites have hitherto been assigned to a separate species of the Coccidia infective to man and designated *Isospora hominis*; this is now invalidated.

The development of the Sarcosporidia in the muscular system of the secondary host also shows some divergence from that of *Toxoplasma*. The interior of the cyst tubes is divided into chambers, and in large tubes only the peripheral chambers contain parasites (figure 64 e). In even larger tubes, amoeboid cells closely pressed against each other can be seen at the periphery, while sickle-shaped zoites which are reminiscent of *Toxoplasma*, but which reach up to 15 µm long, lie further inwards; these, as with *Toxoplasma*, infect the host orally.

SUBCLASS 2: CNIDOSPORIDIA (Doflein 1901)

The Cnidosporidia, the "stinging-spore animalcules", produce spores with one or more coiled "sting" threads, which shoot out in the gut of the new host to anchor the spores and facilitate the penetration of the spore contents into the gut wall. The mode of production and the morphology of the spores vary according to the taxonomic Order.

Order 1: Myxosporidia (Bütschli 1881), figures 65–67

The Myxosporidia, or "slime-spore animalcules", are parasites of fish and develop in body cavities such as the gall bladder and urinary bladder, or intracellularly in the tissues. In the adult form they are large multinucleate amoeboid parasites with pseudopodia (figure 65 a–c). In contrast to the Telosporidia, spore production does not occur at the end of the development of the individual; in the multinucleate parasite membranes form around individual large nuclei and the protoplasm surrounding them. These spore enclosures (sporonts or sporoblasts) are therefore situated intracellularly in the parasite. Nuclear divisions and the production of further internal membranes (figure 66 a, b) lead to the production within the pansporoblasts of predominantly multinucleate and at the same time multicellular spores (figure 66 b). Figure 67 illustrates diagrammatically the whole cycle of the genus **Myxobolus**. In the uninucleate diploid zygote (a), nuclear division (b) begins with meiosis (c) in which 2 residual nuclei remain lying in a peripheral position. After further proliferation of the haploid nuclei (d), 2 spores are produced in the pansporoblast (e); each spore has 6 nuclei, of which two become nuclei of the coating structure, two the nuclei of the polar capsule structure, and the two remaining become the nuclei of the amoebula (f). The projectile "sting" threads lie in the two polar capsules (g, figure 66 b). The nucleus of the ameobula reverts to a diploid state by fusion of the two haploid nuclei, whilst the other nuclei are already beginning to degenerate. The diploid amoebula becomes a schizont (k) in a new host. It can also divide further by plasmotomy to some extent (l). In the adult multinucleate parasite (m), individual nuclei can become haploid by meiosis (o–q) and unite with one another (r), thereby producing a zygote endogenously (a) from which the sporont (pansporoblast) develops within the parasite. These spore enclosures are hence intracellular cells in which further intracellular production of cells occurs once again. This type of multicellular structure differs morphologically and physiologically from the simple or multi-layered tissues of the Mesozoa and Metazoa, in particular because the unicellular amoeboid germ forms again at the end, so that classification of the Cnidosporidia with animals having tissues does not appear to be justified. Some Myxosporidia can give rise to economically important diseases in commercially useful fishes, e.g. *Myxobolus pfeifferi* causes boil disease in barbel and *Myxosoma cerebrale* to spin or twist disease in trout.

Order 2: Actinomyxidia (Štolc 1899), figure 68

The Actinomyxidia, the "ray-slime animalcules", are parasites of the gut and body cavities in bristle worms (Oligochaeta), e.g. *Tubifex* and starworms (*Sipunculida*). They are amoeboid multi-nucleate parasites which develop to pansporoblasts directly, and have three-rayed spore formation. Three polar capsules are present in each spore. The spores usually contain eight or more uninucleate amoebulae; only in the genus **Tetractinomyxon**, development of which can be seen from the diagram in figure 68, is a single amoebula formed in each spore. The spore develops from the uninucleate zygote (a). When 8 nuclei are produced (b), three of them become exospore nuclei, one the endospore nucleus, three the nuclei of the three polar capsules, and one in the inner region becomes the nucleus of the amoebula (c). This latter divides into 2 unequal nuclei (d). In the new host, the amoebula is released (e) but does not then develop into schizonts, as occurs in the Myxosporidia, but straight into sporoblasts. After the first binary fission of the amoebula (f), the two smaller nuclei become the nuclei of the coat of the pansporoblast (g, h). The two cells within the young pansporoblast divide, although not at the same time (i), until each has produced 8 daughter cells (k), which become haploid gametes by discarding a residual body from their nucleus (k) and unite (l, m) to produce 8 diploid zygotes (a).

Order 3: Microsporidia (Balbiani 1882), figures 69, 70

The Microsporidia, or "small-spored animalcules", are common intracellular parasites of arthropods and fishes. One, 2, 4, 8 or more spores arise in each pansporoblast. Apart from the amoebula, each spore contains only a vacuole with the polar thread. The small size of the parasite has occasionally led to divergent descriptions of its development. Figure 69 shows as an example the development of *Stempellia magna* from mosquito larvae of the genus *Culex*. The zygote becomes a sporoblast from which sometimes only one spore develops (b, c, g). However, after a single nuclear division in stage (d), 2 spores are formed (e, f) and, by further division during stage (d), sometimes 4 or 8 spores can be produced. The vacuole of the ripe spore contains a coiled polar thread (h) which can be everted in the gut of a new host (i). The emerging amoebula (k) passes through the gut wall and travels by way of the blood into the fat bodies of the mosquito larva (l) where, after 4-nucleate (q) or multinucleate (s) stages, binucleate sporonts are finally constricted off (r, t); their nuclei undergo karyogamy (u, v).

The development of *Nosema bombycis*, the causal agent of the serious spot disease of silkworms, takes a similar course (figure 70 a). The uninucleate stage in the muscular system of the host gives rise to elongated forms with many nuclei which break up into binucleate pieces which then proceed to form spores. In the Microsporidia of fish, too, e.g. in *Glugea anomala* in the stickleback, corresponding reproductive stages are seen in the greatly enlarged cells of the connective tissue until, finally, usually only sporoblasts and perfect spores are to be found in the cyst-like enlarged cells of the host (figure 70 b).

SUBCLASS 3: HAOLOSPORIDIA (Caullery and Mesnil 1899) figures 71, 72

The Haplosporidia, the "simple-spore animalcules", constitute a heterogeneous group of Sporidia in which several species are even reminiscent of the lower fungi. Their development has by no means been made clear. In contrast to the other Sporozoa, the sporogony here plainly proceeds without previous gamogony, as shown by the development of *Ichthyosporidium giganteum* from the wrasse *Crenilabrus melops* (figure 71). The amoebula (a) arising from the spore penetrates from the gut into various organs and becomes multinucleate by schizogony (b). Several multinucleate plasmodia are formed from this by plasmotomy (c, d), and their nuclei lie together in pairs (e, f) and then lose their nuclear membrane (g); this procedure does not correspond to the behaviour of a protozoan cell and is reminiscent of the naked nuclei of bacteria. Sporogony begins in the course of plasmotomy through disintegration into individual pieces with paired nuclei (h, i). The sporont produced in this way proceeds through further divisions (k–n) to the formation of spores (o).

Pneumocystis carinii (figure 72) is of interest in human medicine as the causal agent of interstitial pneumonia in infants; it blocks the bronchioles and alveolae of the lungs and leads to death by asphyxia. The amoebula (figure 72 a) released from the "spores" surrounds itself with a 7–10 μm coat of mucus and reproduces by binary fission (b–e). During sporogony the parasite, which is otherwise 1–2 μm, enlarges within the gelatinous coating so that the latter is almost filled (f–h). Eight spores are formed in each of these sporoblasts (i). This is the phase at which microscopic diagnosis is possible. Here, too, a truly protozoan nature is questionable, since *Pneumocystis* sporoblasts show gram-positive staining as found in fungi. Experimental transmission has not yet been successful, despite the fact that the parasite, which always lives extracellularly, has been demonstrated in various animal species, particularly rodents.

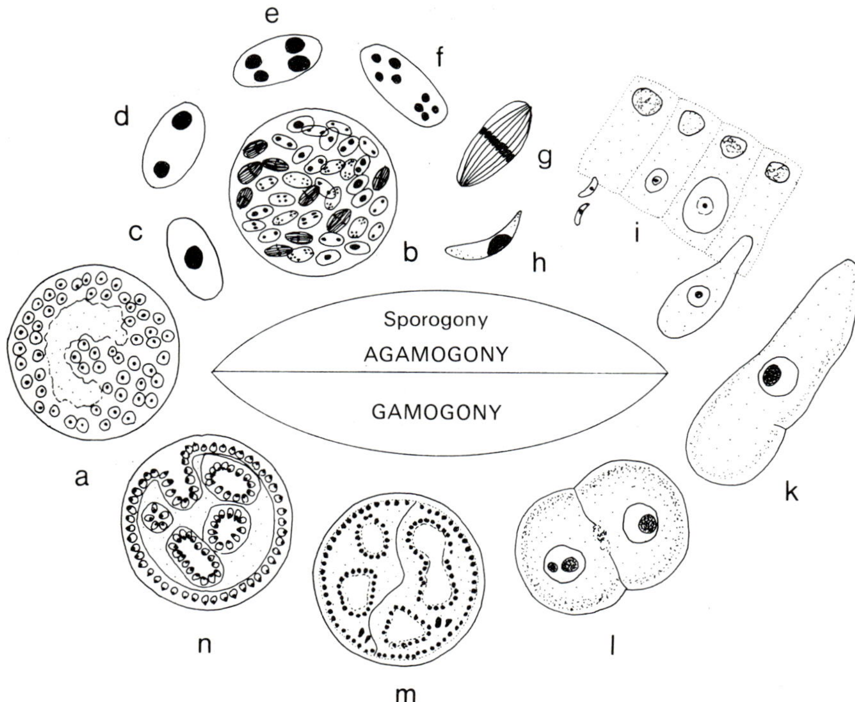

Figure 46. **Eugregarinida:** cycle of *Lankesteria culicis*
Explanations in the text (p. 93)

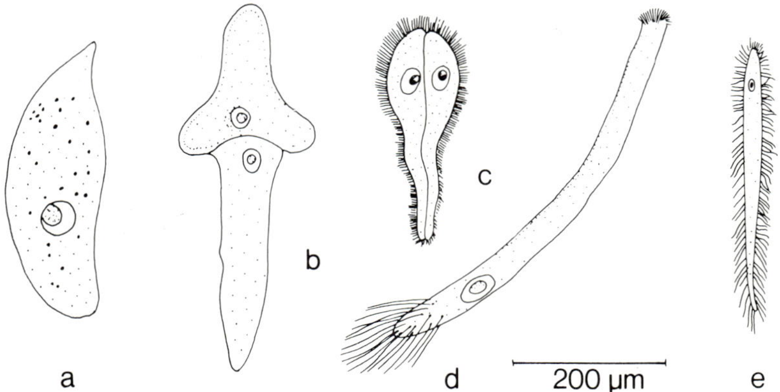

Figure 47. **Eugregarinida, Monocystidae** (Magnification 105 ×)
  a *Monocystis rostrata* from the earthworm *Lumbricus terrestris*
  b *Enterocystis ensis* (syzygy) from the may fly *Caenis* sp.
  c *Urospora chiridotae* (syzygy) from the holothurian *Chiridota laevis*
  d *Nematocystis vermicularis* from *Lumbricus terrestris*
  e *Rhynchocystis pilosa* from *Lumbricus terrestris*

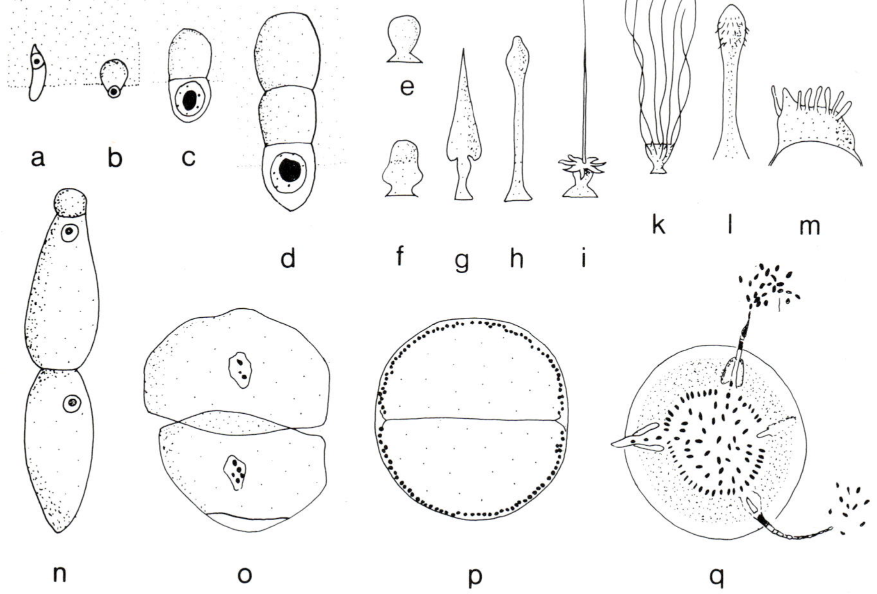

Figure 48. **Eugregarinida, Polycystidae**
a–d early stages of *Gregarina cuneata* from the mealworm (larva of *Tenebrio molitor*)
e–m epimerites of:
   e *Gregarina longa*
   f *Sycia inopinata*
   g *Pileocephalus heeri*
   h *Stylocephalus longicollis*
   i *Beloides firmus*
   k *Cometoides crinitus*
   l *Gemorhynchus monnieri*
   m *Echinomera hispida*
n–q gamogony of *Gregarina ovata* from the earwig *Forficula auricularia*
   n syzygy
   o onset of encystment
   p peripheral gamete production
   q sporangium with expulsion of spores through the sporoducts

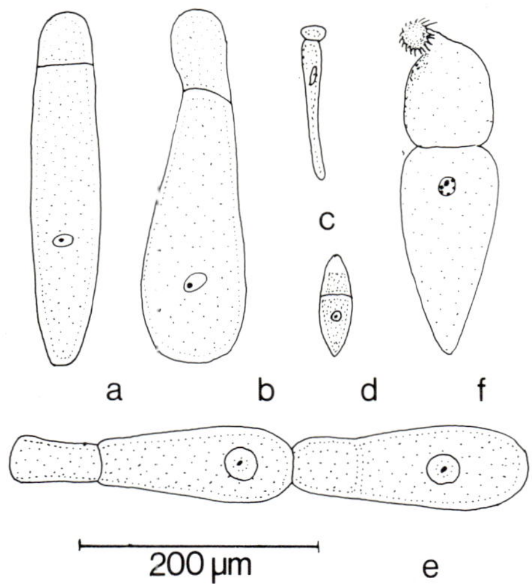

Figure 49. **Eugregarinida, Polycystidae** (Magnification 140 ×)
a–e Gregarinida of the mealworm (*Tenebrio molitor*):
  a *Gregarina polymorpha*
  b *G. cuneata*
  c *G. steini*
  d *Steinina ovalis*
  e syzygy of *G. cuneata*
  f *Actinocephalus dujardini* (with epimerite) from the centipede *Lithobius forficatus*

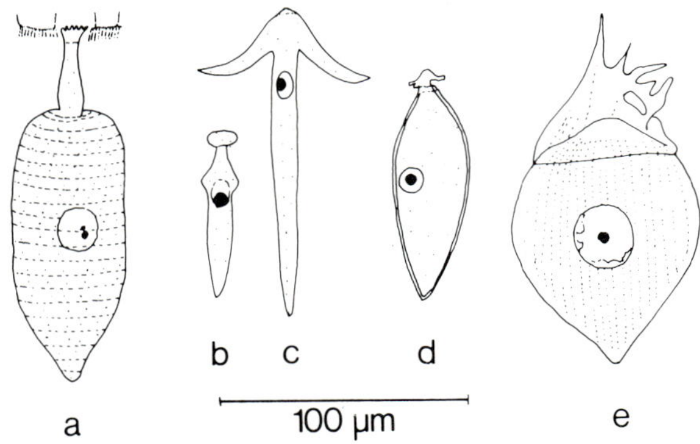

Figure 50. **Eugregarinida, Polycystidae** (Magnification 275 ×)
a *Lecythion thalassemae* (with epimerite) from the annelid *Thalassema neptuni*
b–c *Ancora sagittata* from the polychaete *Capitella capitata* (**b** immature form with epimerite)
d *Polyrhabdina spionis* (with epimerite) from the polychaete *Scololepsis fuligionosa*
e *Hentschelia thalassemae* (with epimerite) from the annelid *Thalassema neptuni*

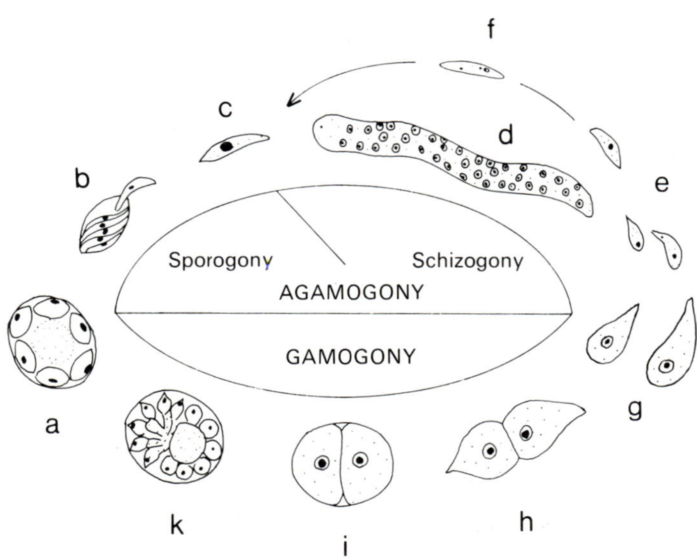

Figure 51. **Schizogregarinida**
Cycle of *Schizocystis gregarinoides* from the gut of the Diptera larva *Ceratopogon solstitialis*. Explanations in text (p. 95)

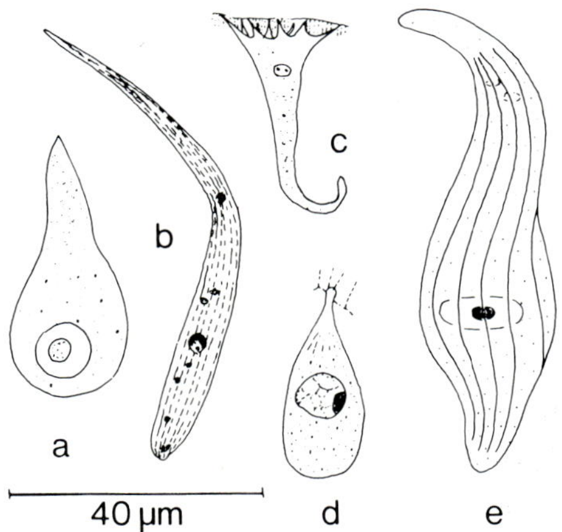

Figure 52. **Schizogregarinida** (Magnification 690 ×)
   a *Caulleryella pipientis* from the gnat larva of *Culex pipiens*
   b *Machadoella triatomae* from the predatory bug *Triatoma dimidiata*
   c *Ophryocystis schneideri* in the Malpighian vessels of Tenebrionida
   d *Merogregarina amaroucii* from the ascidian *Amaroucium* sp.
   e *Selenidium potamillae* from the polychaete *Potamilla reniformis*

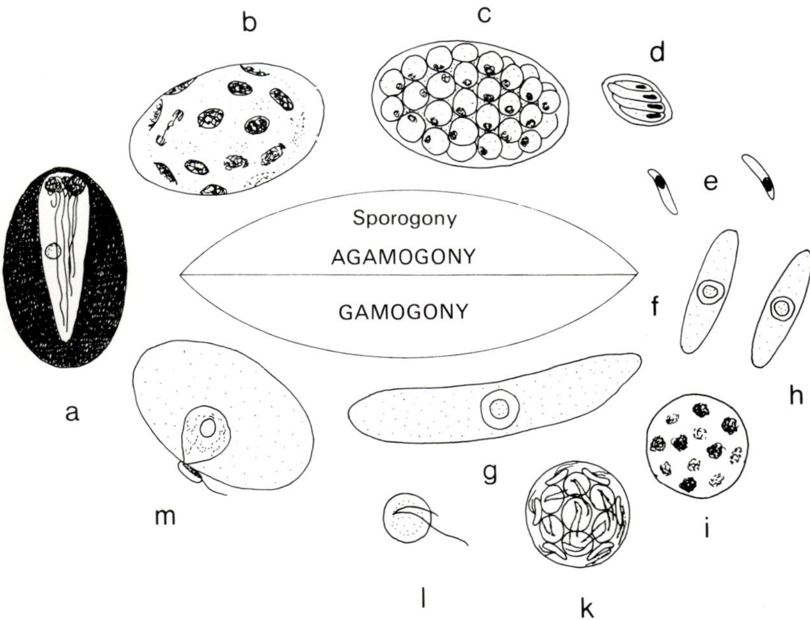

Figure 53. **Protococcidia**
Cycle of *Eucoccidium dinophili* from the archiannelid *Dinophilus gyrociliatus*. Explanations in the text (p. 96)

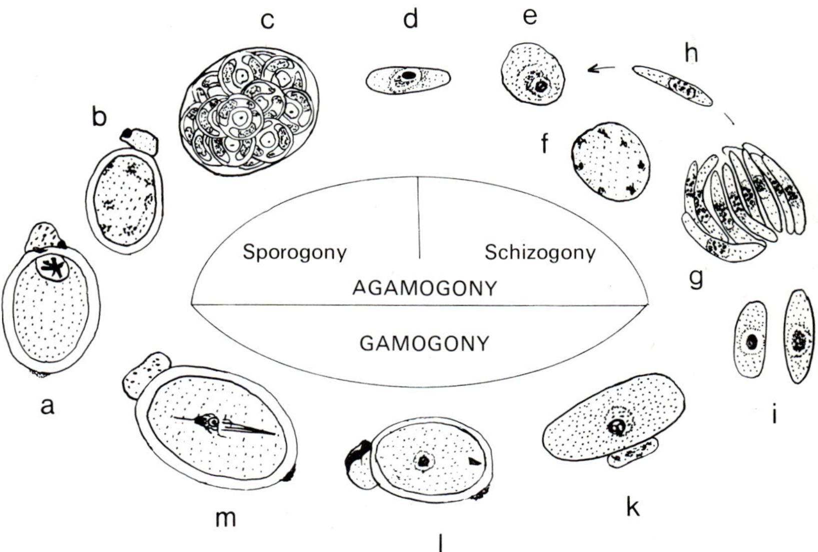

Figure 54. **Schizococcidia, Adeleidea**
Cycle of *Adelea ovata* from the centipede *Lithobius forficatus*. Without alternation of hosts. Explanations in text (p. 97)

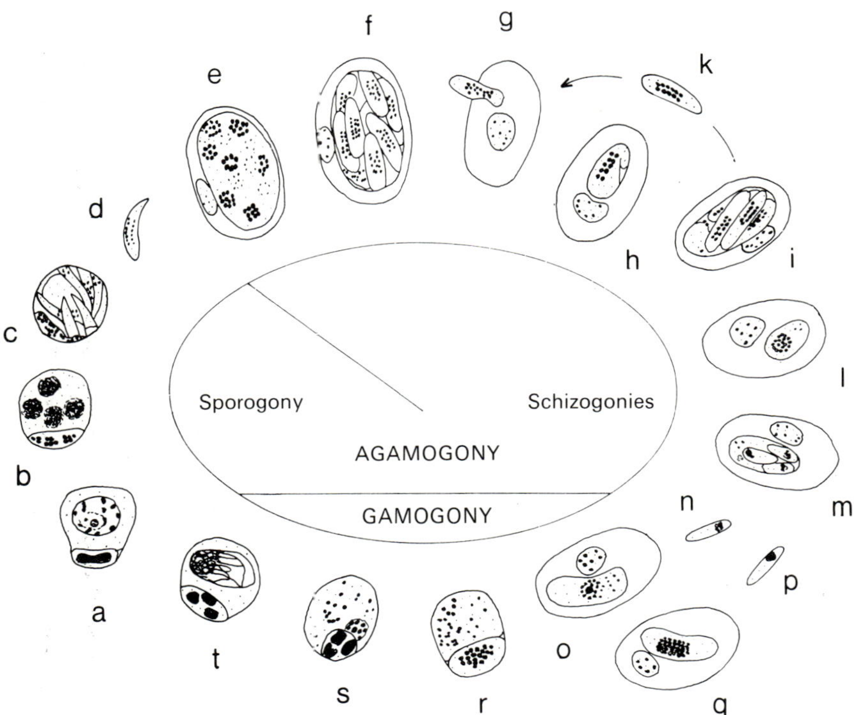

Figure 55. **Schizococcidia, Adeleidea**

Cycle of *Haemogregarina stepanowi*.

d–o, q in the turtle *Emys orbicularis*

o, q–d in the leech *Placobdella catenigera* (alternation of hosts). Further explanations in the text. The host cells in the figures e–i, l, m, o, q each contain their own nucleus as well as the parasite.

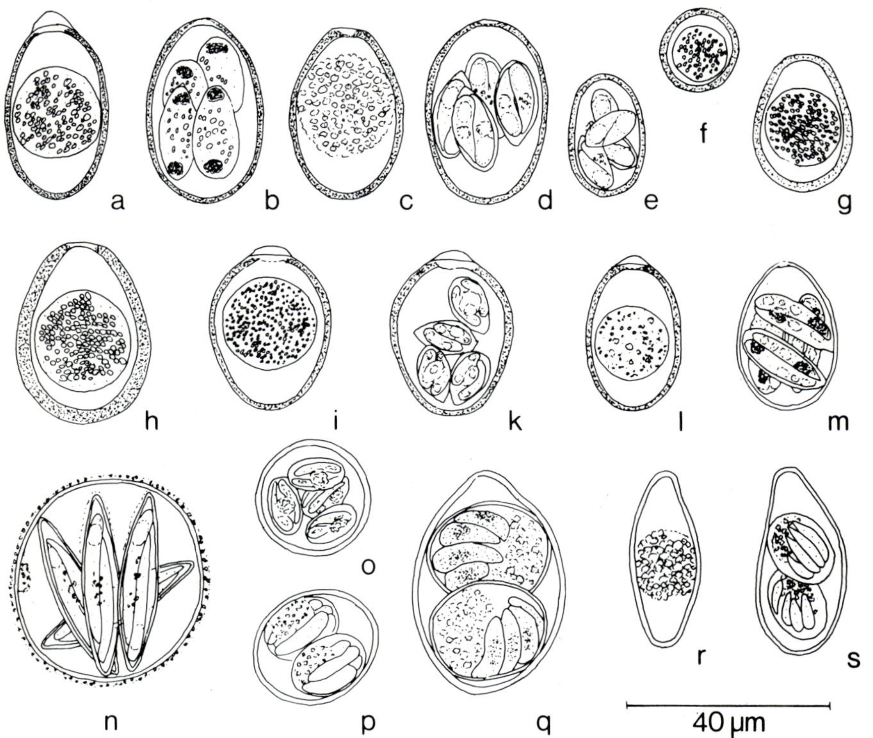
a
b
c
d
e
f
g
h
i
k
l
m
n
o
p
q
r
s
40 μm

Figure 56. **Schizococcidia,** oocysts of various **Eimeridea** (Magnification 690 ×)
a–b *Eimeria stiedae* (rabbit liver)
  a unripe
  b with 4 unripe spores
c–d *E. magna* (rabbit gut)
  c unripe
  d ripe with 4 spores each with 2 sporozoites
  e *E. perforans* (rabbit gut), ripe
  f *E. zürni* (cattle), unripe
  g *E. bovis* (cattle), unripe
  h *E. bukidnonensis* (cattle), unripe
i–k *E. granulosa* (sheep)
  i unripe
  k ripe
  l *E. arloingi* (sheep), unripe
  m *E. debliecki* (pig), ripe
  n *E. sardinae* (in testicles of sardines, sprats and herrings), ripe
  o *E. clupearum* (in the liver of herrings, mackerel and sprats), ripe
  p *Isospora rivolta* (dog), ripe with 2 spores each with 4 sporozoites
  q *I. felis* (cat), ripe
r–s *I. belli* (man)
  r unripe
  s ripe

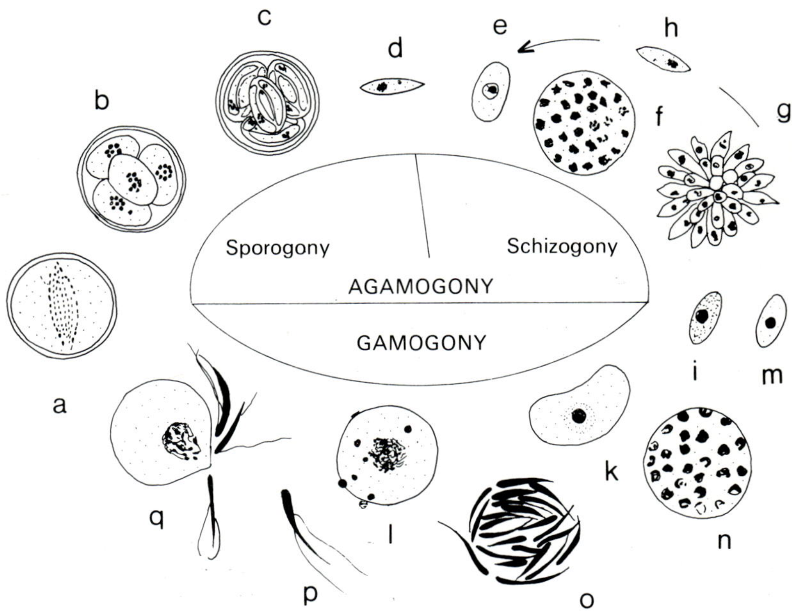

Figure 57. **Schizococcidia, Eimeridea**
Cycle of *Eimeria schubergi* from the centipede *Lithobius forficatus*. Without alternation of hosts. Explanations in the text (p. 98)

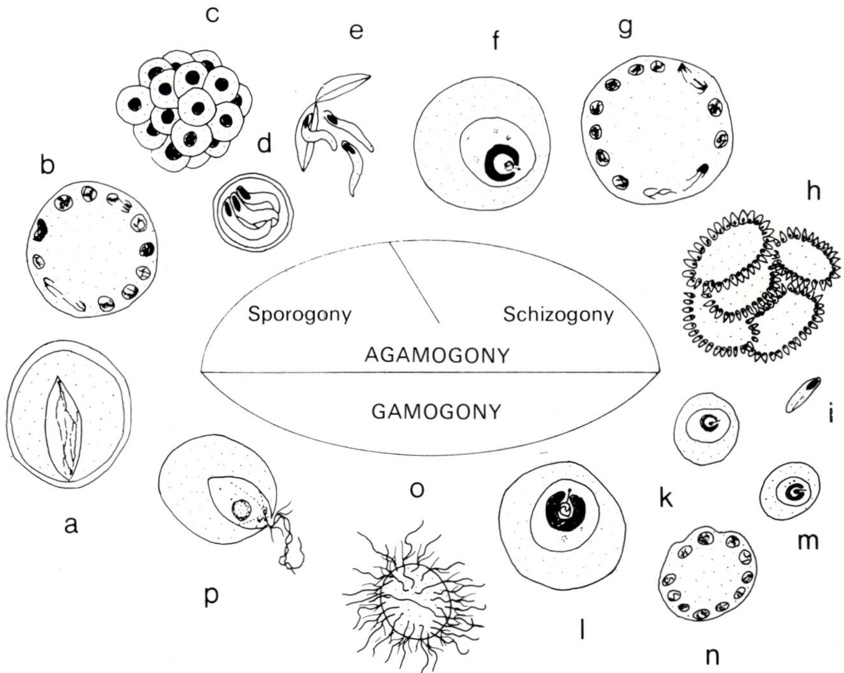

Figure 58. **Schizococcidia, Eimeridea**
Cycle of *Aggregata eberthi*
   e–h in the crab *Portunus depurator*
   i–d in the cuttlefish *Sepia officinalis* (alternation of hosts). Further explanations in the text (p. 99)

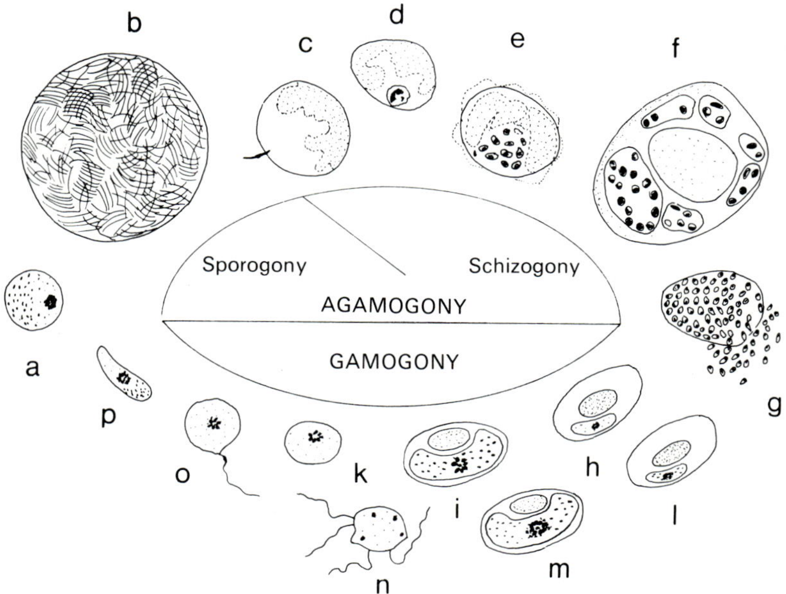

Figure 59. **Schizococcidia, Haemosporidia**
Cycle of *Haemoproteus columbae*
   c–i in the pigeon *Columba livia*
   k, n–b in the louse-fly *Lynchia maura* (alternation of hosts). Further explanations in the text (p. 100)

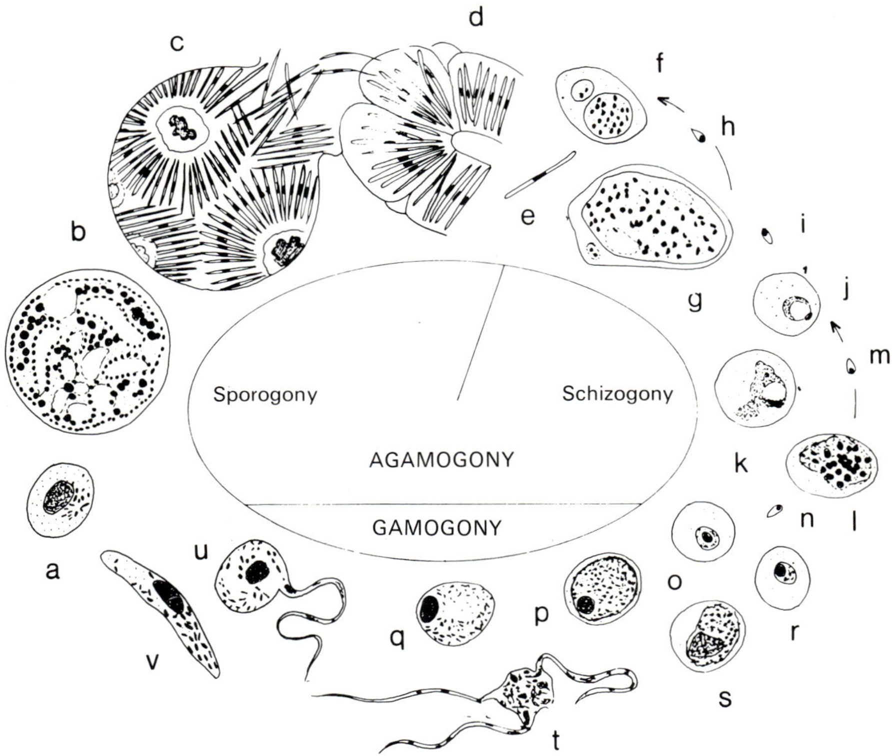

Figure 60. **Schizococcidia, Haemosporidia**
Cycle of *Plasmodium vivax* (causal agent of vivax **malaria**)
   e–p, s in man
   q, t–d in the *Anopheles* mosquito (alternation of hosts). Further explanations in the text (p. 100)

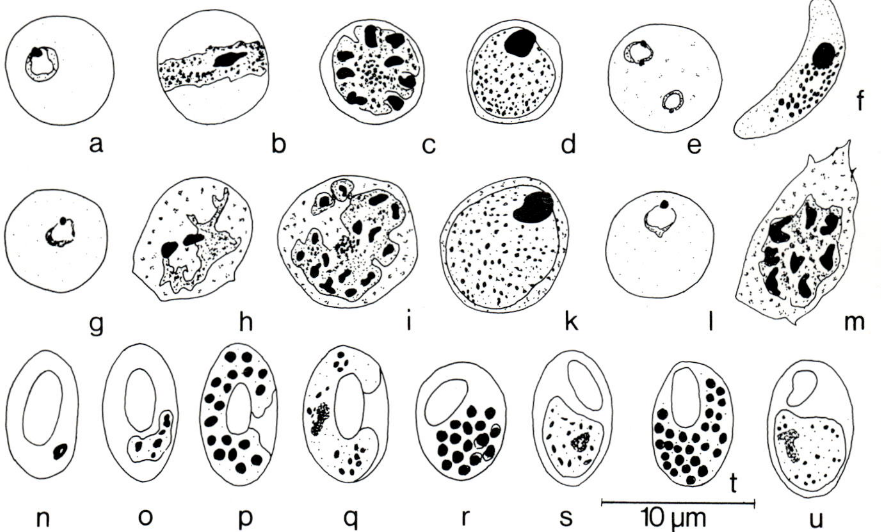

Figure 61. **Schizococcidia, Haemosporidia** (Magnification 2060 ×)
Malaria parasites in erythrocytes.
a–m in man
a–d *Plasmodium malariae* (causal agent of quartan malaria)
  a young uninucleate ring form in erythrocyte
  b young schizont (band form)
  c mature schizont with usually 8 merozoites
  d macrogamont
e–f *Plasmodium falciparum* (malignant tertian malaria)
  e young ring forms, uninucleate and binucleate
  f free macrogamont (erythrocyte disintegrated)
g–k *Plasmodium vivax* (benign tertian malaria)
  g uninucleate ring form
  h onset of schizogony
  i mature schizont with usually 16 merozoites
  k macrogamont
  h–k erythrocyte enlarged with Schüffner's dots
l–m *Plasmodium ovale* (West African form of tertian malaria)
  l uninucleate ring form
  m mature schizont with usually 8 merozoites, erythrocyte oval, enlarged
  with Schüffner's dots
n–u avian malaria in nucleate erythrocytes
n–q *Plasmodium circumflexum* (blackbird and other birds)
  n uninucleate ring form
  o young schizont
  p mature schizont (with 13–30 merozoites)
  q macrogamont
r–s *Plasmodium cathemerium* (sparrow and other birds)
  r schizont (with 6–24 merozoites)
  s macrogamont
t–u *Plasmodium gallinaceum* (in the Indian domestic fowl)
  t schizont (with 20–36 merozoites)
  u macrogamont. All microgamonts have a larger, more diffuse nucleus than
  the macrogamonts.

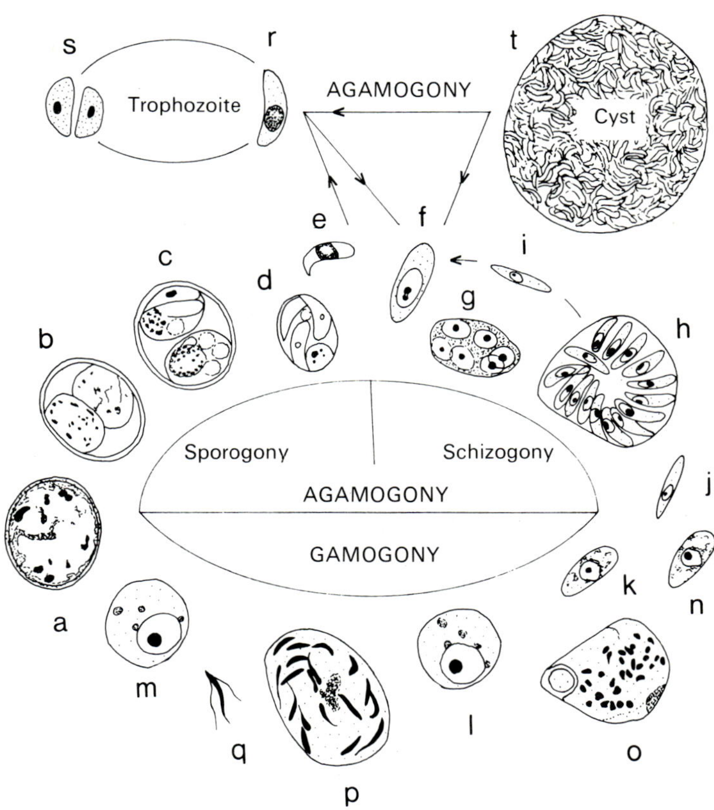

Figure 62. **Schizococcidia, Toxoplasmida**
Cycle of *Toxoplasma gondii*
  a–q in the gut of the cat (primary host)
  r–t in secondary host. Further explanation in the text (pp. 101, 102).

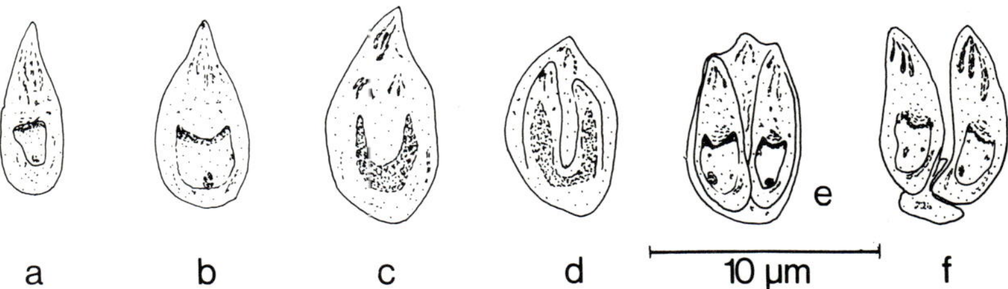

Figure 63. **Endogeny** in *Toxoplasma gondii*, development of 2 daughter cells within the mother cell. Further explanations in the text (p. 102).

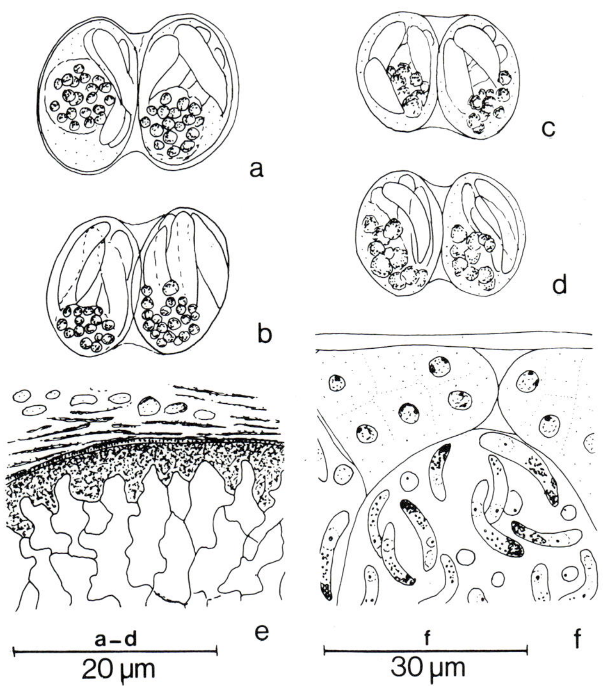

Figure 64. **Toxoplasmida, Sarcosporidia**
a–d oocysts (Magnification 1375 ×)
a–b *Sarcocystis fusiformis*, oocysts with 2 spores each with 4 sporozoites
  a from the dog
  b from man
  c *Sarcocystis tenella*, oocyst with spores from the cat
  d *S. miescheriana*, oocyst from man
  e part of a utricle of *Sarcocystis fusiformis* in the musculature of cattle (low
magnification), peripheral chambers containing parasites, inner chambers
void
  f part of a utricle of *Sarcocystis tenella* in the oesophageal musculature of
the sheep. Amoeboid, uninucleate stages towards the outside, sickle-shaped
zoites towards the centre (Magnification 920 ×)

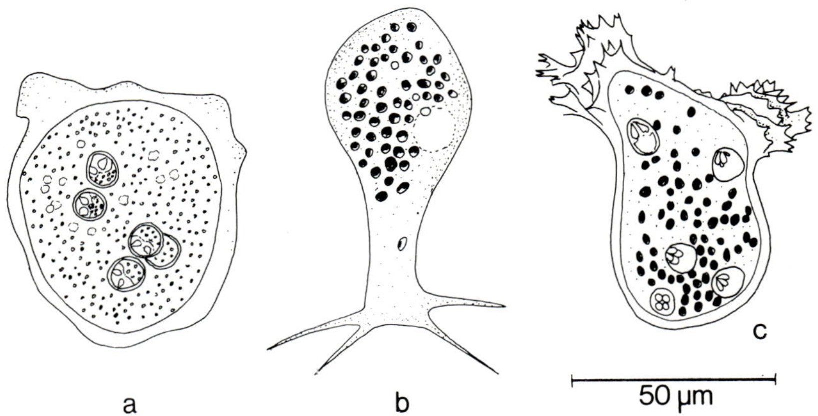

Figure 65. **Myxosporidia** (Magnification 550 ×)
a *Sphaerospora divergens* from the renal canalicula of the slime fish *Blennius pholis* and the lip fish *Crenilabrus melops* and *C. pavo.* 5 spores in the endoplasm
b *Leptotheca agilis* from the gall bladder of the thornback *Trygon pastinaca.* Unstained, with a round pansporoblast
c *Chloromyxum leydigi* from the gall bladder of various selachians. 5 spores in the endoplasm

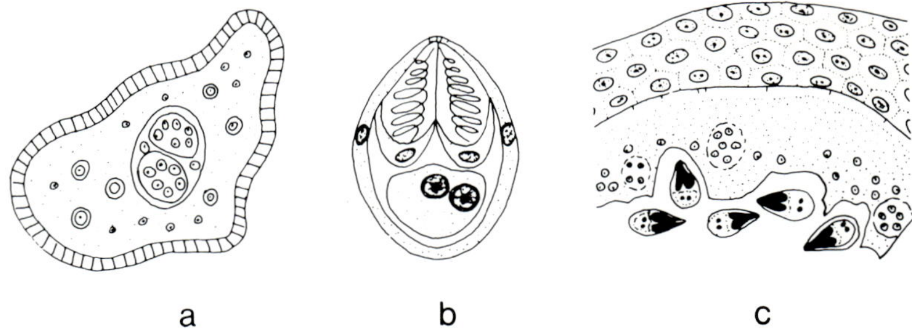

Figure 66. **Myxosporidia**
  a diagram of **Myxobolus**: ectoplasm striated, vegetative nuclei in the endoplasm, 6 endogenous cell structures = beginning of the pansporoblasts, 1
pansporoblast with 2 spore structures
  b mature spore of the genus *Myxobolus*: amoebula with 2 nuclei, 2 polar
capsules at the front with coiled threads
  c *Myxosoma dujardini* in the gill lamellae of sardines, river perch, roach and
carp. Sector of a cyst; ectoplasm under the gill epidermis, then endoplasm
with vegetative nuclei and 3 pansporoblasts. Mature spores in the lumen.

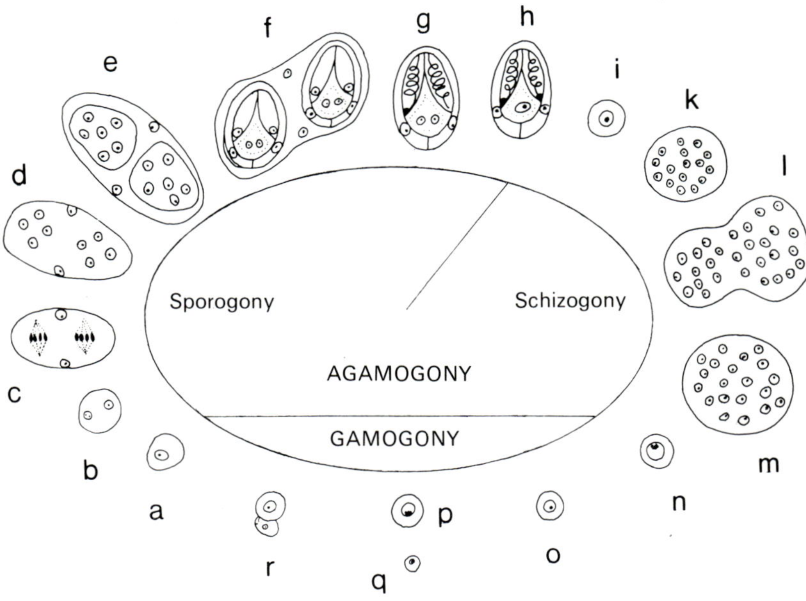

Figure 67. **Myxosporidia**
Cycle of the genus **Myxobolus**. Explanations in the text (p. 104).

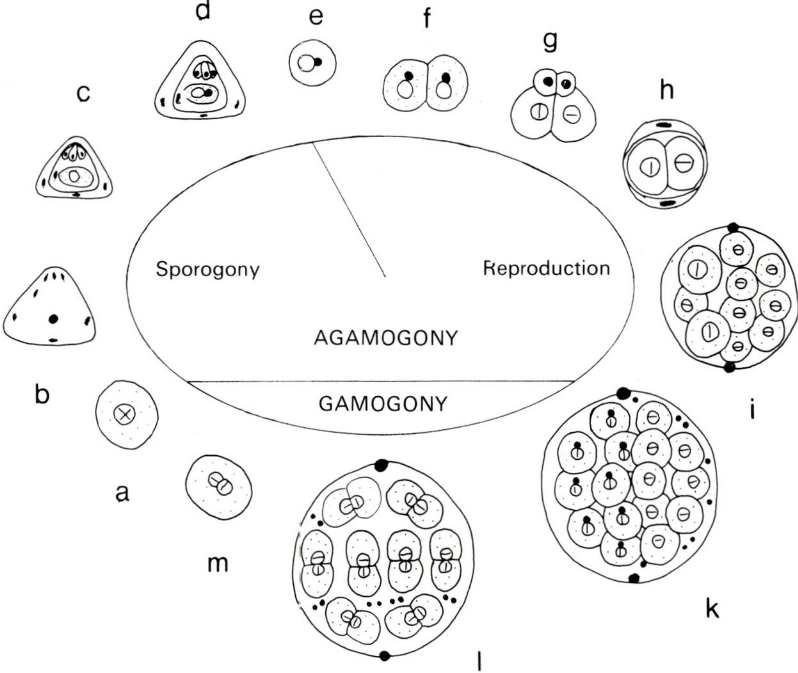

Figure 68. **Actinomyxidia**
Cycle of *Tetractinomyxon intermedium* from the body cavity of the sipunculid (starworm) *Petalostoma minutum*. Further explanations in the text (p. 105).

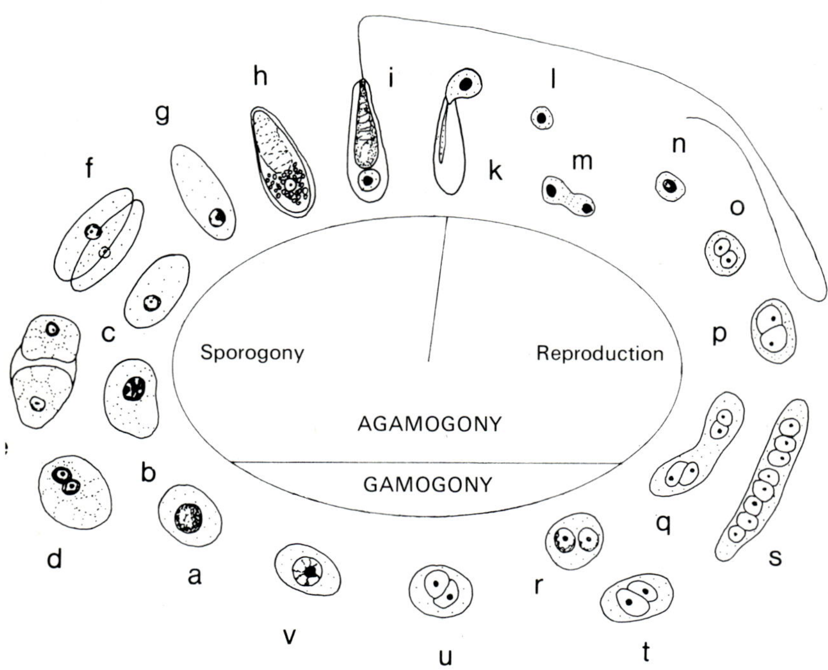

Figure 69. **Microsporidia**
Cycle of *Stempellia magna* from the fat bodies of the gnat larva of the genus *Culex*. Explanations in the text (p. 106).

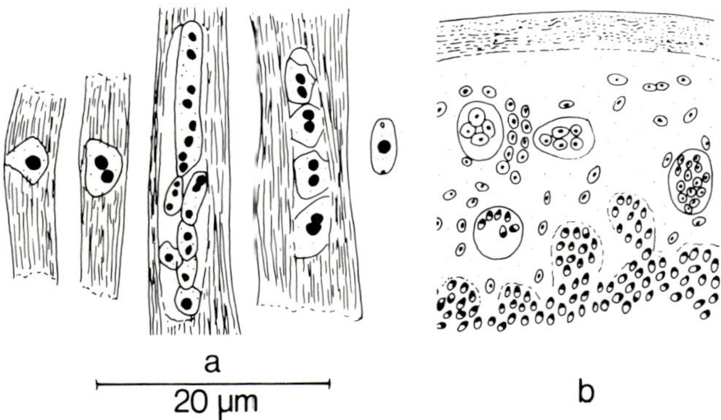

Figure 70. **Microsporidia**
  a *Nosema bombycis* in the muscles of the silkworm (Magnification 1375 ×)
Development to a multinucleate plasmodium, dissociation into binucleate
pieces, from which the uninucleate spores are formed.
  b *Glugea anomala* from the stickleback. Sector of a multinucleate host cell
in the subcutaneous connective tissue, much enlarged by infection, with
pansporoblasts and mature spores of the parasite.

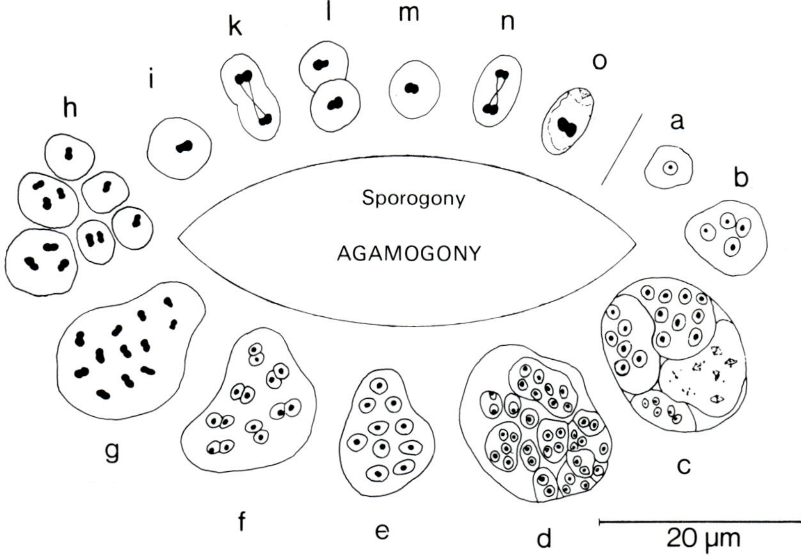

Figure 71. **Haplosporidia**

Cycle of *Ichthyosporidium giganteum* from the wrasse *Crenilabrus melops.* Explanations in the text (p. 106) (Magnification 1375 ×)

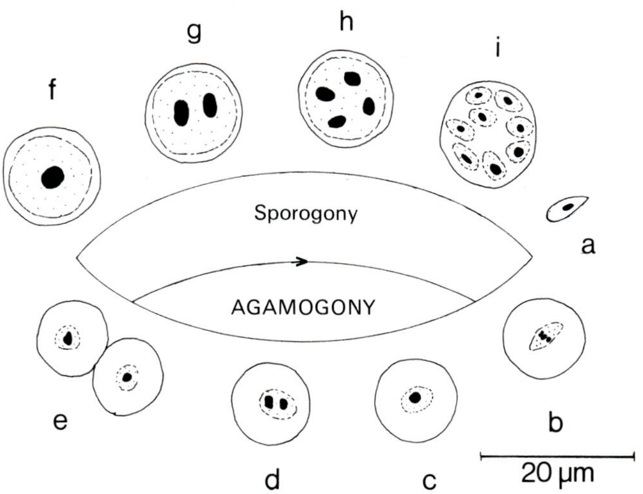

Figure 72. **Haplosporidia**

Pneumocystis carinii from the lungs of a human infant. Explanations in the text (p. 107) (Magnification 1030 ×)

CLASS IV: CILIATA (PERTY 1852), CILIATED ANIMALCULES OR CILIOPHORA (DOFLEIN 1901), CILIA BEARERS.

THE CILIATA ARE ALSO CALLED INFUSORIA OR INFUSION ANIMAL-cules since some of them tend to develop in, for example, hay steepings. They are, like the Flagellata, characterized by the possession of pulsating locomotory organelles. The cilia or little hairs are shorter than flagella in terms of their size relative to that of the cell body. This difference is only one of degree, and even electron-microscopic structure analyses of the locomotory organelles have not revealed a fundamental difference between the two. Nuclear dimorphism, the presence of micronucleus and macronucleus, was therefore considered specific for the Ciliata. However, the occurrence of macronuclei in the Foraminifera shows that the formation of a macronucleus is not a specific characteristic of the Ciliata. Although most of the Ciliata differ from other Classes in having a highly developed polyploid macronucleus, other Ciliata have only diploid macronuclei like those of the Foraminifera, and some species are multinucleate without nuclear dimorphism, as are many of the Polymastigina. This latter group form the Protociliata, in antithesis to the heterokaryotic Euciliata. The homokaryotic Protociliata are today included as Flagellata in the Polymastigina by some authors who ignore the original definition referring to cilia. However, some ciliates found in the interstices of sea sand have, like the Protociliata, no macronucleus but can be assigned to the Euciliata on the basis of the rest of their cell structure; because of this, the problem of the homokaryotic ciliates is not solved by inclusion of the Protociliata with the Flagellata.

SUBCLASS 1: HOMOKARYOTICA, PROTOCILIATA (Metcalf 1918), figure 73

Metcalf described as Protociliata, or "primitive ciliated animals", the family Opalinidae, which are predominantly gut parasites of tailless amphibians, but sometimes occurring also in amphibians with tails, in some reptiles and in fish. Their cell surface is entirely ciliated, holotrichous, with the cilia arranged in diagonal rows, and more dense at the front than at the rear. These parasites do not have a cell mouth. The Opalinidae are binucleate or multinucleate, and have a cylindrical or flattened cell body. These characteristics lead to the separation of 4 genera (see figure 73 a–d). The genus **Stephanopogon** is also homokaryotic with similar nuclei and a nucleolus. This genus does not correspond to the standard type recognized for the Protociliata but, within the Euciliata, to the tribe Prostomata of the holotrichous Gymnostomata (figure 73 e) on account of its anteriorly placed cell mouth and trichites.

SUBCLASS 2: HETEROKARYOTICA, EUCILIATA (Metcalf 1918), figures 74–91

The Euciliata, the "proper, true" ciliates, all show nuclear dimorphism. In most cases the macronucleus is polyploid, but in a few species is diploid. All the latter species, with the exception of those of the genus **Loxodes**, live in the mesopsammon, the interstitial system of sea sand, as the genus **Stephanopogon** (lacking a macronucleus) does. Within the heterokaryotic Euciliata, however, it is not the nuclear structure which is the critical taxonomic criterion, but primarily the nature of the cilia. The location and form of the cell mouth also determine systematic position.

Order 1: Holotricha (Stein 1859), figures 74–81

The Holotricha, the "completely hairy" ciliates, can, with their uniform mantle of cilia, be regarded as the basic form for all Euciliata; from them are derived morphological variations, such as the restriction of cilia to certain areas of the surface, and the cementing of cilia together into smaller or larger sail-like surfaces, the membranellae and membranes (figure 123 f) or into styloid locomotory organelles, the cirri (figures 86, 123 g). Also, as mentioned before, all of the species without a macronucleus or with only a diploid macronucleus are holotrichous. The comprehensive order, the Holotricha, comprises the suborders Gymnostomata, Trichostomata, Hymenostomata, Thigmotricha, Astomata and Apostomea.

The **Gymnostomata**, the "naked-mouthed" ciliates (figures 74, 75), have a superficial cell mouth (cytostome) which leads into a cell gullet (cytopharynx). The cytopharynx may contain toxicysts, which secrete a toxin to paralyze prey, or rigid to paralyze rod-shaped trichites, which surround the gullet as a basket structure to strengthen it mechanically and thereby help to break up the food (figure 74 a, g, h). Species with rounded cell bodies and anterior cell mouths are combined together as the tribe **Prostomata** (figures 74 a–d, 75 a–e). The laterally flattened **Pleurostomata** have their cytostome on the narrow ventral side (figures 74 e–f, 75 f–g), the similarly flattened **Hypostomata** on the broad flattened ventral side (figure 74 g–h).

The cytostome structure of the **Trichostomata** (figure 76) is not on the cell surface but at the bottom of an indentation (the oral cavity or peristome), which is characterized by a particular yet simple ciliature for wafting the foodstuff inwards; this is reflected in the name of the suborder. The **Hymenostomata** (figures 77, 78, 94 e–g), like the Trichostomata, have an oral cavity in which, however, the cilia do not function as individuals but are partially or completely fused into membranelike structures known as **membranellae**. The **Thigmotricha** (figure 79) are marine ciliates, which live parasitically, predominantly in the mantle cavity of mussels. Towards the anterior end on the left side are a number of tactile-sensitive cilia, a thigmotactic area of cilia, which commences ciliatory beating when touched by the host organism, and consequently leads to the attachment of the parasite. The cytostome structure lies in the posterior half of the cell, or else a proboscis is formed at the front end with which the parasite anchors itself in the host.

The **Astomata**, the "mouth-less" ciliates (figure 80), are also parasites. They live predominantly in the gut lumen or in the body cavity, and occasionally also as tissue parasites in the organs, of oligochaetes. The genus **Haptophyra** is an exception in that it lives in amphibia. The completely ciliated cell has no cytostome. Many species produce a sucking depression for attachment to their host, and some produce an outer skeleton. An intracellular skeleton may also occur. Multiplication proceeds by terminal budding (figure 96a).

The **Apostomea** (figure 81), which are parasites mainly of marine crabs, have cilia in rows and arranged in a clockwise spiral, with a ventral mouth area which is specially ciliated. The Apostomea undergo metamorphosis. As tomites (figures 81b, 96d) they are free-living ciliates, but when they fasten on to their host they produce cyst coats to become phoronts; from these the vegetative parasitic stage, the **trophont**, emerges (figures 81a, c, 96b). If multiplication begins in an adult trophont (figure 96c), it is then called a **tomont**. The terminal ciliated buds which detach themselves during this process give rise again to the free-living forms, the tomites (figure 96d).

Order 2: Spirotricha (Bütschli 1889), figures 82–86

The Spirotricha, the "twisted-haired" ciliates, have a series of membranellae oriented in a clockwise direction and leading from the anterior pole to the cytostome; this is the adoral zone of membranellae. The individual membranellae are transverse platelets formed by cilia adherent together. The Spirotricha comprise the suborders Heterotricha, Oligotricha, Entodiniomorpha, Odontostomata and Hypotricha.

The **Heterotricha** (figure 82, figure 1c–d), "variously-haired" ciliates, have, apart from the strongly developed adoral zone of membranellae, a uniform ciliature usually covering the whole roundish cell. The **Oligotricha** (figure 84), the "few-haired" ciliates, are, with the exception of the adoral zone of membranellae, aciliate or have only a few cilia or stiff bristles. The peristome field always lies at the truncated anterior end. The **Entodiniomorpha** (figure 83), named after the genus *Entodinium* ("whirler within"), are gut parasites of herbivorous mammals. They develop in large numbers in places where cellulose is fermented, in the rumen and reticulum of ruminants, and in the caecum and colon of horses. The retractable adoral zone which is wound in a clockwise direction and is made up of cilia joined together into cirri, generates a strong vortex for feeding; proper membranellae are formed only deeper in the oral cone. Apart from the additional zones of cirri present in some species, the pellicle is aciliate as in the Oligotricha, and furthermore is strengthened by internal fibrillar structures and is often provided with lobe-like and thorn-like processes. Special fibrillar lattice structures of the cell body, the skeletal plates, serve at the same time as the chief site for carbohydrate storage. Although large food particles [of the host] are often washed in, the ciliates do not participate in the breakdown of cellulose but are dependent on the starch content of the food. The **Odontostomata** or **Ctenostomata** ("tooth- or comb-mouths") (figure 85) are small laterally-flattened ciliates with long cilia, which are arranged in a few rows or groups, particularly at the posterior end. The adoral zone, usually restricted to 8 membranellae, lies in a capsular cavity opening ventrally. The pellicle, with armour-like strengthening, often has thorn-like processes at the back. These are ciliates of putrescent freshwater mud containing hydrogen sulphide. The **Hypotricha** (figure 86) ("hairy-below") are ciliated only on the ventral side, and here the cilia are all or mostly adherent together to form cirri. These flattened spirotrichs have only some tactile bristles remaining on the dorsal side. The peristome, too, with the adoral membranellae, is on the ventral side.

Order 3: Peritricha (Stein 1859), figures 87–89

The Peritricha ("hairy all round") have an adoral ciliated zone consisting of two parallel rows of cilia which run anticlockwise round the peristome area which lies on the truncated anterior end. The zone leads as an open spiral to a vestibulum, a funnel-shaped or tube-shaped peristome cavity situated in front of the true cytostome. The outer row of cilia of the adoral zone is fused into a membrane. The peritrichs are chiefly ciliates which are attached by means of the posterior end or by a special stalk (suborder **Sessilia**). Only a few species have secondarily reverted to free-swimming forms. A region for mobile attachment to the substratum is produced in the family Urceolariidae (suborder **Mobilia**).

The Mobilia (figure 87) live chiefly on the body surfaces, including the gills, of fresh-water and marine animals (polyps, molluscs, vortex worms, crabs and fishes) but sometimes also in internal body cavities (urinary bladder). Their broad aboral end, with a sucker-like ciliated attachment disc, glides over the host. A firm ring of hooks embedded in it supplies internal strengthening, and thereby makes possible the production of partial vacuum when the sucking bell arches outwards. The Sessilia (figures 88, 89) may be stalked or unstalked, and live on water animals and water plants, as well as on dead matter. The ciliature is confined to the adoral zone, surrounded by an annular swelling, the peristome rim, which can close by the contraction of the peristome opening. A temporary wreath of cilia is formed aborally only with the change into the swarmer form, which separates from the stalk when a new location is to be found. The attached, often bell-shaped, peritrichs, live either singly, solitarily or, after incomplete division, as a colony. Many species are strongly contractile as a result of the development of myonemes. While most of the Sessilia have no casing or only a gelatinous coat (tribe **Aloricata**, figure 88) some produce a strong pseudochitinous shell into which they can draw back by contraction (tribe **Loricata**, figure 89).

Order 4: Chonotricha (Wallengren 1896), figure 90

The Chonotricha, the "hairy-funnelled" ciliates, form a small group of ciliates parasitic on crustacea (amphipods). They have an anterior funnel-shaped ectoplasmic peristome, at the base of which parallel rows of cilia run in a clockwise direction to the cytostome. The rest of the cell body is devoid of cilia. Multiplication is achieved by budding with the production of ciliated swarmers.

Order 5: Suctoria (Claparède and Lachmann 1858), figure 91

The Suctoria, or "sucking infusoria", suck dry their prey, mostly ciliates, with the help of tentacles. In the adult state they are aciliate, attached, with or without a stalk, and immobile, so that they are also called **Acineta**. Some species produce a shell. Multiplication is usually by budding, and the detached buds are completely ciliated and, like the holotrichous swarmers, seek out a new location. After attachment, they change once more into the aciliate true suctorial form. They occur in fresh water and in sea water.

Reproduction and Sexuality of the Ciliates

The **agamic reproduction** of the homokaryotic **Protociliata** (figure 73 a–d) proceeds by plasmotomy. The nuclei present are distributed between the two daughter animals produced by a **diagonal division** (figure 92) which follows the direction of the rows of cilia; their number is made up mitotically during later growth. The genus **Stephanopogon** (figure 73 e), also homokaryotic, behaves differently. The cells are binucleate at first but produce 16–24 nuclei by repeated mitotic divisions during growth. Division then proceeds within a thin cyst coat by dissociation into 8–12 binucleate offspring.

The normal agamic reproduction of the heterokaryotic **Euciliata** is by binary fission which takes the form of a **transverse division** (figure 94 a) in contrast to the longitudinal division of the flagellates. Mitotic division of the micronucleus occurs first and, after distribution of the daughter micronuclei between the two halves of the cell, the polyploid macronucleus divides by simple constriction during cell division. The daughter animals add their missing cell halves after the division (figure 94 b–c). In stalked forms, e.g. in the sessile **Peritricha**, division is **longitudinal** with reference to the stalk; in this case the direction of division of the cell body lies transversely to the stalk axis (figure 94 d). The formation of colonies on a common stalk base is thereby made possible. A different mode of reproduction occurs in **Ichthyophthirius** (figure 94 e–g). The parasite which has grown up in the skin of a fish encysts and, in the free cyst, division gives rise to two daughter cells, which then produce many small **swarmer forms** by numerous nuclear divisions. In some ciliates numerous macronuclei lie singly or in chains. In these cases, all of the macronuclei clump together before division. After the subsequent elongation of the now single macronucleus, agamic separation of the cell and macronucleus follows (figure 94 h–k).

Unlike the "secondary" polyploid type of macronucleus, the "primary" diploid macronuclei, in genera such as **Loxodes**, **Geleia** and **Remanella** (cf. Holotricha, Gymnostomata, figures 74 f, 75 f), are unable to divide. In this way they correspond to the diploid macronuclei of the Foraminifera. In such cases the macronuclei are first distributed between the two daughter animals during the agamic transverse division. After division of the micronuclei, the balance of macronuclei is made up from the micronuclei formed (figure 93 a–d). In the genus **Trachelo-raphis**, which is also holotrichous, the micronuclei and macronuclei assemble together to form a complex nucleus, which in *T. phoenicopterus* is formed from 6 micronuclei and 12 macronuclei (figure 93 e). During division, a mitosis of the micronuclei is followed by distribution of the macronuclei to the two halves of the nuclear complex (figure 93 f) in which a further mitotic division of the micronuclei is then followed by the formation of 6 new macronuclei from 6 micronuclei (figure 93 g–h).

If the daughter cells arising from a division are very unequal in size, the division is called **budding**. In the holotrichous **Astomata** (figure 80), terminal budding is widespread (figure 96 a). In this, a significantly smaller individual with micronucleus and macronucleus is cut off at the posterior end; often, complete separation is not achieved. Chain formation arises through further budding before the buds are finally cut off. Terminal bud formation also occurs in some holotrichous **Apostomea**. The alternation of generations and of forms of these parasitic ciliates have already been considered in the taxonomic section (page 138). In the genus **Chromidina**, the slender vermiform parasitic trophont produces numerous terminal buds (figure 96 b, c) which eventually become detached as tomites and form the free-living generation (figure 96 d). The **Chonotricha** (figure 90) also produce buds. In these attached forms, the free-swimming buds or swarmers, have the function of finding a new habitat. Figure 90 b shows a bud still attached to the mother cell in the genus **Chilodochona**. In *Spirochona gemmipara*, which is shown in figure 96 e, the difference between the bud and the stationary mother cell is not only one of size. The free bud (ciliated in an inner space which is open at the side) does not yet possess the peristome funnel typical of the adult stage (figure 96 f). At the posterior end of the lateral split in the swarmer is the site of the future basal disc with which the swarmer will attach itself to the host.

Budding is widespread and varied in the **Suctoria,** e.g. in *Dendrosomides paguri* buds are produced singly, and this leads only to the detachment of an arm (figure 91 h, left arm) which then becomes re-attached by the production of a stalk. The genus **Sphaerophrya** (figure 95 a–c) also begins by producing single buds in which binary fission usually results in unequal daughter cells which can either be ciliated swarmers or aciliate. Other species of Suctoria always produce several buds at the same time. *Ephelota gemmipara* provides an example of multiple external budding (figures 91 m, 95 d). As in each division, a complete nuclear apparatus must be produced in each daughter cell here also; in this case, the branching of the macronucleus is particularly clear. With the production of a ciliated area, the buds are released as swarmers. In the parasitic *Tachyblaston ephelotensis* (figure 95 e–o), a suctorian with alternation of generations, the developmental phase parasitic in the head of *Ephelota gemmipara* undergoes multiple external budding with production of swarmers (figure 95 h, i). The free-living generation, which is shown here (figure 95 m–o) developing on the stalk of *E. gemmipara,* forms a succession of tentacle-bearing daughter cells (dactylozoites), following which there is total distribution of the old macronucleus and disintegration of the mother cell.

Many Suctoria have no outer (exogenous) budding but form swarmers within the mother cell. This internal (endogenous) budding begins with the formation of a **brood chamber**, an indentation in the cell body which remains connected with the exterior through the birth pore (figure 95 p). Within the brood chamber the bud arises as a protuberance into which the nucleus grows, as it does in exogenous budding. The bud, when developed, is freed as a ciliated swarmer (figure 95 q, r). It attaches itself to a suitable substratum and loses its cilia at the same time as tentacles are formed. In unfavourable environmental conditions, suctoria can also produce swarmers without budding. *Tokophrya cyclopum* (figure 95 s–u) provides an example of this. With the formation of an internal brood chamber, the entire animal changes into a swarmer which seeks a more favourable habitat for itself, leaving behind the empty coat with the stalk of the original suctorian.

The sexual process in the protociliates, like agamic division, differs from that in the euciliates. *Opalina ranarum* provides an example of the sexual process in **protociliates** (figures 73 c). Whilst agamic reproduction by diagonal plasmotomy of the multinucleate ciliates (figure 92 b) is normally the only process to occur, in the early part of the year, when the frogs spawn, small forms with only few nuclei arise through divisions which follow each other in rapid succession. These forms subsequently surround themselves with a cyst coat. The cysts (figure 97 a) are expelled with the faeces, infect young tadpoles and, once in the gut, lose their cyst coats. The *Opalina* cells, which are still small, either grow through nuclear multiplication into normal forms, or undergo repeated divisions to form microgametes and macrogametes which unite with one another (figure 97 b, c). The zygote formed by this anisogamous **fusion** is a cystozygote, which is again expelled with the faeces and infects new tadpoles in which the typical multinucleate *Opalina* cells form after disintegration of the cyst membrane (figures 97 d–g, 73 c).

Whilst the protociliates, like the flagellates and all other Protozoa undergo syngamy as the sexual process, fertilization in the **euciliates** takes place by **conjugation**, i.e. mutual fertilization of two sexual partners after a characteristic process for production of gametic nuclei. This process is described in more detail in figure 98 using *Paramecium caudatum* as the example (figure 77 b). Initially the conjugation partners, or conjugants, lie side by side as a conjugation pair (figure 98 a). Each conjugant still has the normal complement of nuclei, which in *P. caudatum* is 1 micronucleus and 1 macronucleus per cell. The diploid micronucleus forms 2 diploid micronuclei (b) in each conjugant at the first pregamic division. Four haploid micronuclei are formed at (c) the second pregamic division, which is a reduction division or meiosis. Three of these four micronuclei degenerate, whilst the remaining one carries out the postmeiotic division (d). Of the two haploid nuclei now present in each conjugant, one crosses over as gametic nucleus (migratory nucleus) into the other conjugation partner, while the second gametic nucleus (stationary nucleus) remains in its original conjugant (e). In each conjugant the two nuclei fuse into a diploid zygotic nucleus (karyogamy, synkaryon formation) (f). The conjugants part again and hence become exconjugants. The disintegration of the old macronucleus, which has already begun, now becomes more obvious. The zygotic nucleus undergoes the first postzygotic division (g). Eight diploid micronuclei (h, i) arise in each exconjugant through two further postzygotic divisions. Four of these micronuclei become four new macronuclei after the old macronucleus has disintegrated. Three micronuclei degenerate, so that now only one micronucleus remains (k). After a mitotic division of the remaining micronucleus, there is an agamic division of the exconjugants with distribution of the nuclei now present (l). After repeated division of the micronucleus, with subsequent cell division, the normal nuclear apparatus of 1 micronucleus and 1 macronucleus per cell is again restored (m). The whole process in Paramecium, including the reorganization of the nuclear complement, lasts up to 4 days.

Conjugation in other euciliates proceeds basically along the same lines, although the number of micronuclear mitoses can vary according to species. Figure 99 summarizes another conjugation process, this time in *Chilodonella uncinata* (compare figure 74 h). Here, too, each conjugant normally has 1 micronucleus and 1 macronucleus (figure 99 a). In pairing, the conjugants coalesce at the mouth region. After the mitotic division of the micronucleus, each partner contains 2 diploid micronuclei, each with 4 chromosomes (b). Following meiosis, 3 micronuclei in each conjugant degenerate, together with the macronucleus. One haploid micronucleus with only 2 chromosomes remains (c). After this divides, the migratory nuclei are exchanged (d), karyogamy occurs and each exconjugant then contains 1 diploid micronucleus, again with 4 chromosomes (e). Two micronuclei arise in each exconjugant (f) by the postzygotic mitotic division of the micronucleus, and one of these develops into a new macronucleus (g) to restore the normal nuclear complement.

Despite certain variations in the course of conjugation, all conjugations in the euciliates can be reduced to a simple **basic formula**. Whether or not there are normally several micronuclei or only one, only one haploid micronucleus remains in each conjugant and both gametic nuclei are formed from this (pronuclei = migratory nucleus and stationary nucleus). After karyogamy, the whole nuclear apparatus of the exconjugant is always formed from a single diploid synkaryon whilst the old macro-nucleus disintegrates. The conjugation is always a gamontogamy, since the formation of pairs is accomplished by the mother cells of the two gametic nuclei.

The outward appearance of conjugation can be very varied. Since in the examples cited the conjugation partners were of equal size, such a conjugation is known as **isogamous conjugation** or isogamonty. If microgamonts and macrogamonts can be distinguished (figure 99 h), this is known as **anisogamous conjugation** or anisogamonty. Anisogamous conjugation is to be seen particularly in attached ciliates, e.g. in the peritrichous vorticellids (figure 100 a–c). In this case the microconjugant is set free and seeks out the attached macroconjugant (a), usually after the microconjugant has undergone one pregamic division more than the macroconjugant. A further peculiarity of this conjugation is that synkaryon formation occurs only in macroconjugants, although both partners produce 2 gametic nuclei beforehand. The total cell content of the microconjugant is absorbed by the macroconjugant; only the empty coat of the microconjugant remains outside and disintegrates (b, c).

This **total cell fusion**, which presents an appearance of syngamy apart from the peculiar mode of production of the gametic nuclei, also occurs in some other species, e.g. in the hypotrichous genus **Uròstyla** (see figure 86 e). Figure 100 d–f shows the fusion of conjugants in the multinucleate species *Urostyla polymicronucleata*. In the **Chonotricha** which, too, are attached forms, one conjugant is also absorbed by the other. *Spirochona gemmipara* (figures 90 a, 100 g–i) illustrates the search for a neighbouring partner, and the detachment of one conjugant which is taken up by the sessile conjugant.

Whilst reciprocal fertilization takes place in most **Suctoria**, some also undergo total cell fusion. These include *Ephelota gemmipara* (figure 91 m) and *Tokophrya cyclopum* (figures 95 s, 100 k–m). In these cases, cytoplasmic protrusions are formed as paired swellings or conjugation bridges, and make contact with the partner which is also attached to the substratum (figure 100 k). Figure 100 l–m shows two neighbouring conjugants of *Tokophrya cyclopum* after contact has been made, and after the detachment of one conjugant and fusion; synkaryon formation has already occurred. The disintegration of the two macro-nuclei has yet to take place.

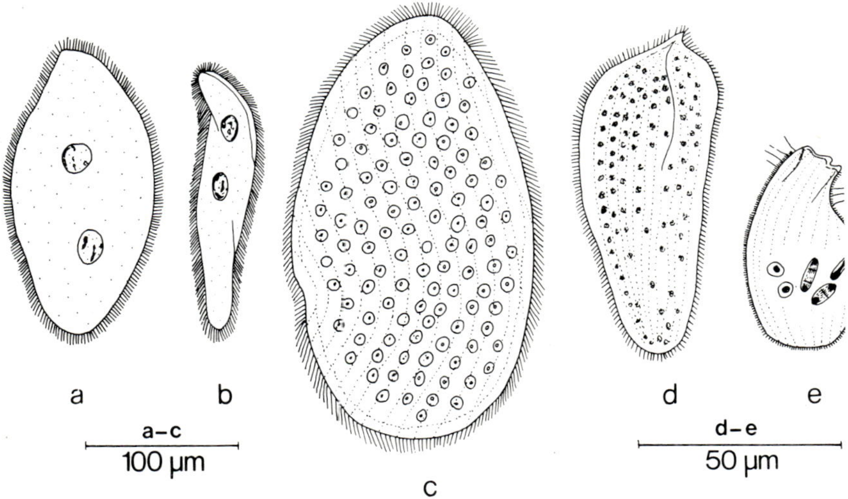

Figure 73. **Homokaryotic ciliates lacking a macronucleus**
a–c (Magnification 205 ×)
d–e (Magnification 550 ×)
a–d **Protociliata**
   a *Zelleriella elliptica* from the toad *Bufo valliceps*, binucleate, flattened cell
   body
   b *Protoopalina intestinalis* from the orange-speckled toad *Bombinator igneus*,
   binucleate, cylindrical
   c *Opalina ranarum* from the grass frog *Rana temporaria*, multinucleate,
   flattened
   d *Cepedea punjabensis* from *Bufo melanostictus*, multinucleate, cylindrical
   e *Stephanopogon mesnili* (**Holotricha**, Gymnostomata [Prostomata]) in the
   interstices of sea sand (mesopsammon). Some of the nuclei in mitosis.

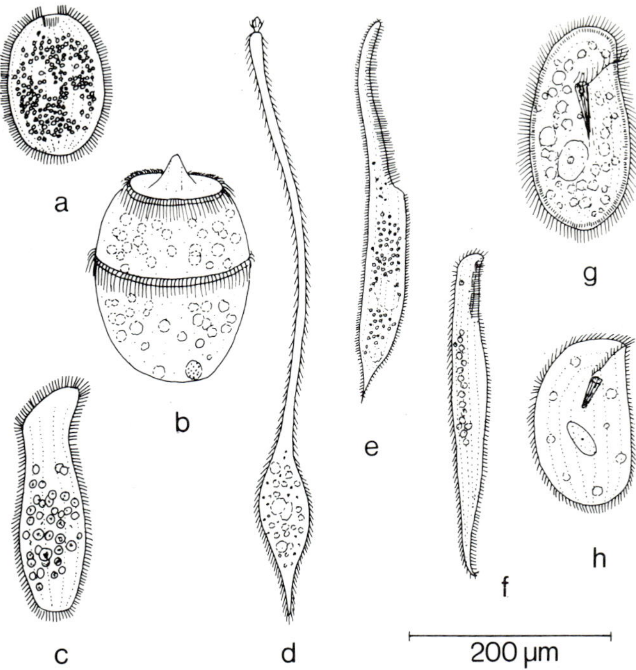

Figure 74. **Holotricha, Gymnostomata** (Magnification 140 ×)
a–d **Prostomata**, Cytostome towards the front
　a *Prorodon ovum*, cytostome with basket apparatus, often green because of zoochlorella, among plants in fresh water
　b *Didinium nasutum*, with 2 belts of cilia, vigorous predator, in fresh water
　c *Spathidium spathula*, with wide anterior mouth opening, fresh water
　d *Lacrimaria olor*, cytostome with trichites on a long contractile neck; among plants in fresh water
e–f **Pleurostomata**, cytostome on the narrow ventral side
　e *Dileptus anser*, proboscis with ventral reinforced row of cilia, cytostome below this with trichocysts. In fresh water, often on plants
　f *Remanella multinucleata*, beak-like anterior end usually with excretory vacuoles (Müller bodies), with needle-shaped internal skeleton, macronuclei diploid; in marine sandy soil
g–h **Hypostomata**, cytostome on broad flat ventral side
　g *Nassula elegans*, cytostome with strong basket apparatus, a reinforced row of cilia at the mouth, cytoplasm variously pigmented through food vacuoles; in standing fresh water
　h *Chilodonella cucullulus*, the same as *N. elegans*, with straight basket apparatus, cytoplasm hyaline particularly at the front; in fresh water usually among filamentous algae

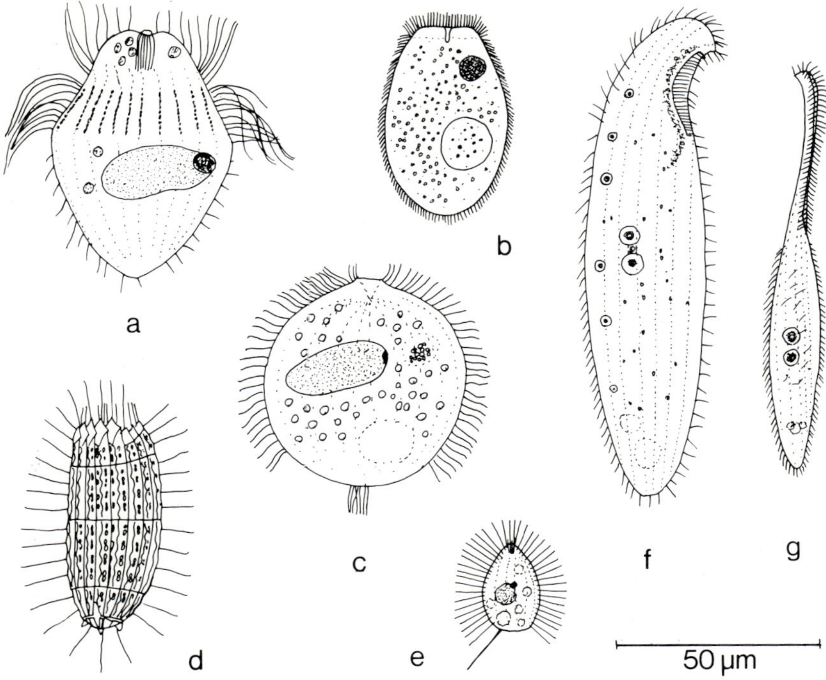

Figure 75. **Holotricha, Gymnostomata** (Magnification 550 ×)
a–e **Prostomata**
a *Askenasia faurei*, like *Didinium*, cytostome strengthened by rods; in fresh water
b *Bütschlia parva*, cilia round the cytostome stronger at the front, frequent parasite in the rumen of ruminants
c *Blepharosphaera intestinalis*, with a tuft of cilia at the back, in the caecum and colon of horses
d *Coleps hirtus*, cell covered with ektoplasmic armour plates, wide cytostome at the front; in putrescent fresh water, frequent among algae
e *Urotricha farcta*, oblique tactile bristle at the back, motion usually circular, then by intermittent jumps. In putrescent water, often among plants
f–g **Pleurostomata**
f *Loxodes rostrum*, with an anterior beak usually with excretory vacuoles, cytostome with peristome-like groove. 2 diploid macronuclei. Preying on algae on the surface of stagnant waters
g *Lionotus fasciola*, proboscis ciliate on one side only, with ventral cytostome and trichocysts; in fresh water frequently among algae.

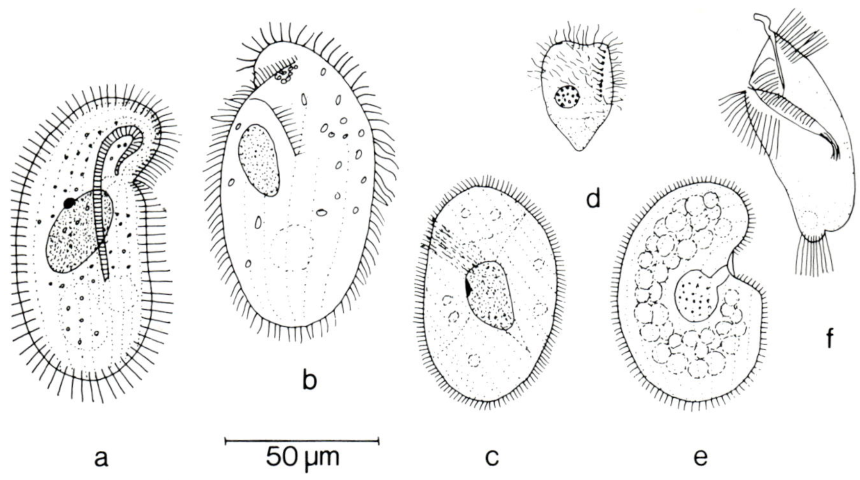

Figure 76. **Holotricha, Trichostomata** (Magnification 410 ×)

a *Plagiopyla nasuta*, flattened dorsoventrally with long ventral peristome groove as far as the depression of the ciliate gullet. In fresh water and the surface of mud

b *Paraisotricha colpoidea*, ventral oral cavity, strong row of cilia in front of this, cell with 19–20 rows of cilia; in caecum of the horse

c *Isotricha prostoma*, densely ciliate all round, cytostome ventral and inclined towards the back, large macronucleus, in the rumen of ruminants

d *Cyathodinium piriforme*, large sunken "peristome cleft" without mouth opening (pseudoperistome), sparse ciliature; in the caecum of guinea pigs

e *Colpoda cucullus*, cytostome funnel-shaped with tunnel-like entrance; in stagnant water, frequently in hay infusions

f *Blepharocoris equi*, oral cavity as a deep ventral furrow with cilia and tentacle-like process, the rest of the ciliature zonal; in the caecum of the horse

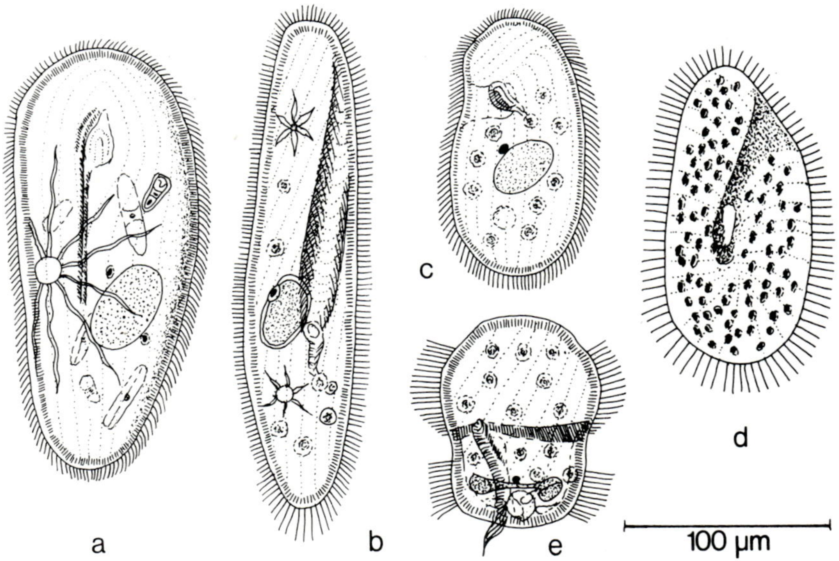

Figure 77. **Holotricha, Hymenostomata** (Magnification 275 ×)
a *Frontonia leucas*, mouth structure anterior with wide vestibule and membrane structures, and aciliate gullet groove adjoining; numerous trichocysts peripheral in the cytoplasm; often with zoochlorella; preys on diatoms. Among plants in mud
b *Paramecium caudatum*, slipper animalcule, cytostome containing a membrane at the base of a deep lateral oral sinus (peristome), trichocysts peripheral in the cytoplasm, cilia elongated at the posterior end of the cell, with 2 contractile vacuoles. Preys on bacteria in marsh waters and infusions
c *Colpidium colpoda*, kidney-shaped, anterior end twisted, vestibule almost three-cornered; preys on bacteria in marsh water and infusions
d *Paramecium bursaria*, shorter and wider than *P. caudatum*, sloped off at the front; nearly always green because of zoochlorella; in fresh water
e *Urocentrum turbo*, barrel-shaped with cilia zones at the front and back, tuft of cilia at the back, long gullet structure at the middle of the cell body, rotary swimming action; in stagnant waters.

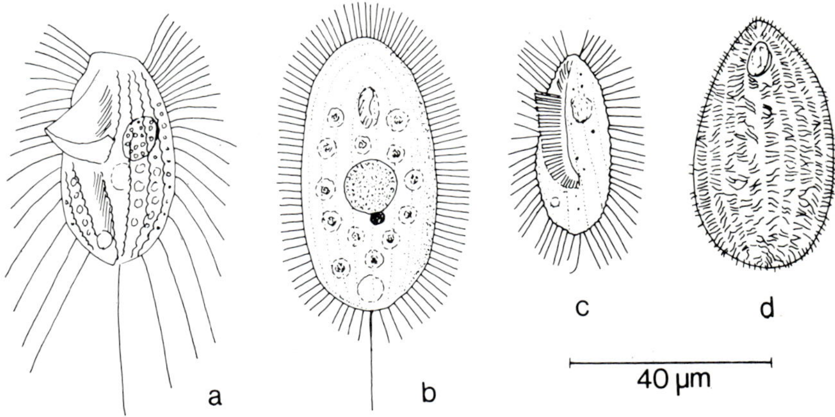

Figure 78. **Holotricha, Hymenostomata** (Magnification 690 ×)
a *Cristigera media*, with sail-like membrane on the peristome towards the front of the cell, long cilia with a trailing bristle; in sea water
b *Uronema marinum*, with powerful trailing bristle, sail membrane only on the right and posterior rim of the vestibule; frequent in stagnant salt and fresh water (infusions)
c *Ctedoctema acanthocrypta*, with long sail-like membrane as far as the end of the cytostome, trichocysts peripheral in the cytoplasm. In fresh water among plants
d *Tetrahymena pyriformis*, with small vestibule with undulating membrane and 3 membranellae, often without a micronucleus; in stagnant fresh water, also found in human faeces.

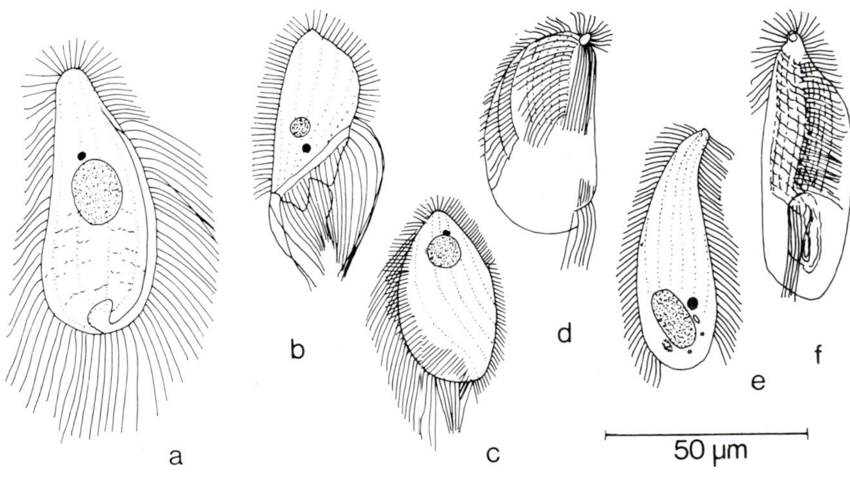

Figure 79. **Holotricha, Thigmotricha** (Magnification 550 ×)

a *Ancistruma japonica*, cilia longer on the right than on the left, cytostome posterior; in the mantle cavity of various marine mussels

b *Plagiospira crinita*, spiral peristome at the middle of the cell body; in the mussels *Cardita calyculata* and *Loripes lacteus*

c *Ancistrina ovata*, with 15–18 rows of cilia parallel to the peristome structure which is marked laterally by two adoral rows of more substantial cilia; in the mantle cavity of *Benedictia limneoides*, etc.

d *Hypocomides mediolariae*, with a sucking proboscis and about 23 rows of cilia; on the gills of *Mediolaria marmorata*

e *Hypocomalgama pholadidis* with sucking proboscis and 25 rows of cilia; in the epithelium of gills and feelers of *Pholadidea penita*

f *Isocomides mytili*, with sucking proboscis, 14–18 rows of cilia at the front; on the gills of the mussel *Mytilus edulis*

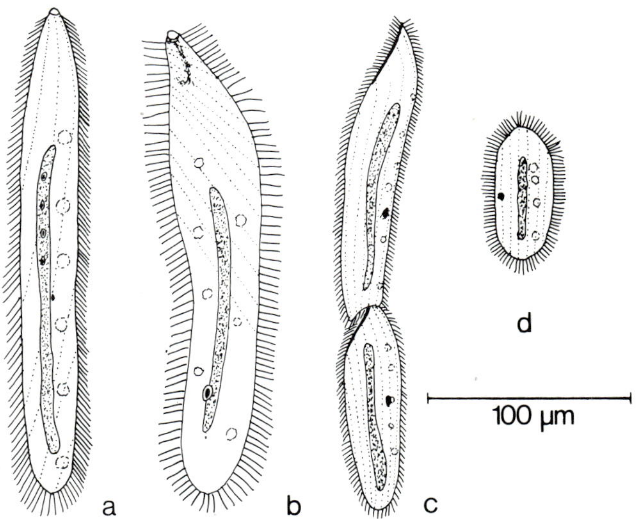

Figure 80. **Holotricha, Astomata** (Magnification 275 ×)
a *Bütschliella ophelia*, ribbon-shaped macronucleus, 10 rows of cilia, anterior aciliate cap; in the polychaete *Ophelia limacina*
b *Intoshellina poljanskyi*, elongated macronucleus, spirally arranged rows of cilia, clasp-shaped skeleton at the anterior end; in gut of the oligochaete *Limnodrilus arenarius*
c *Protoradiophrya fissispiculata*, division stage, laterally at the anterior end with spicula-type skeleton in the cytoplasm; in the front section of the gut of *Styloscolex* sp.
d *Anoplophrya marylandensis*, without skeleton, cilia in longitudinal rows; in the gut of the earthworm *Lumbricus terrestris* and the oligochaete *Helodrilus caliginosus*

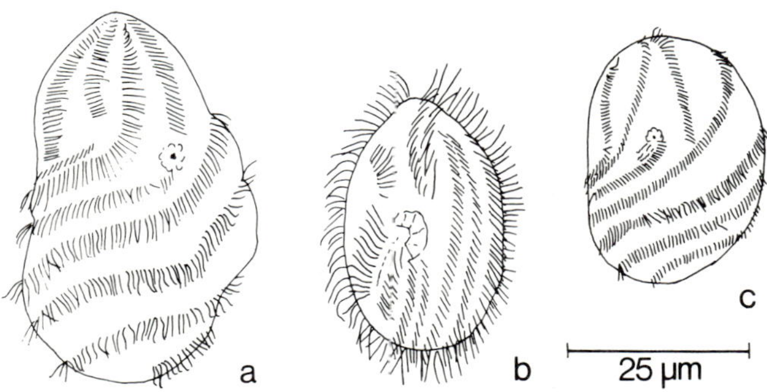

Figure 81. **Holotricha, Apostomea** (Magnification 825 ×)
a *Synophrya hypertrophica*, ventral view of a young trophont, from the discarded crab shell (exuviae) of *Carcinus maenas* and *Portunus* species
b *Spirophrya subparasitica*, ventral view of the tomite with thigmotactile zone of cilia beside the mouth structure. Parasite of the copepod *Idya furcata*
c *Gymnodinioides inkystans*, ventral view of the trophont, among the armour plates of the hermit crab (*Eupagurus*).

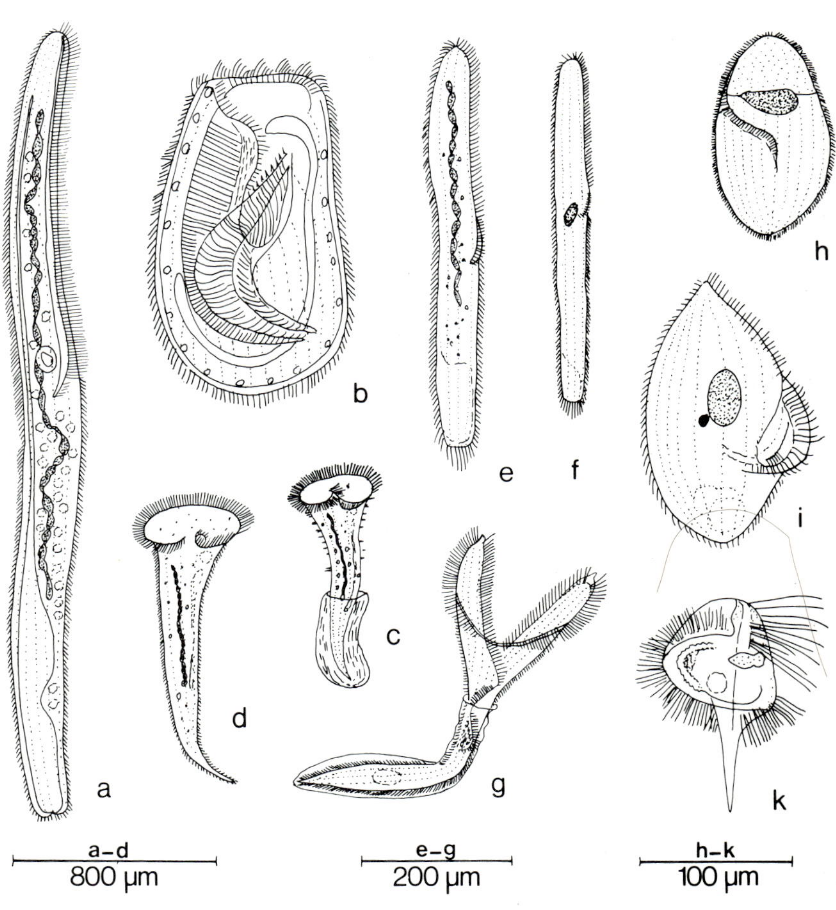

a
b
c
d
e
f
g
h
i
k
a–d
800 μm
e–g
200 μm
h–k
100 μm

Figure 82. **Spirotricha, Heterotricha.** a–d Magnification 34·5 × , e–g Magnification 105 × , h–k Magnification 205 ×

a *Spirostomum ambiguum*, adoral zone of membranellae at the front, surface spirally striated, chain-like macronucleus, cell very contractile; in stagnant marsh water often in large numbers, over 2 mm long

b *Bursaria truncatella*, peristome with zone of membranellae deeply invaginated at the front like a sac, with resupinate oral funnel at the base, can grow to more than 1 mm, in swamp water

c *Stentor roeseli*, wide peristome field at the front with zone of membranellae, can form gelatinous tube when established; rope-like macronucleus; in fresh water, particularly among plants

d *Stentor polymorphus*, usually coloured green by zoochlorellae, chain-like macronucleus, cell contractile; on sticks and plants in fresh water

e *Spirostomum intermedium*, chain-like macronucleus

f *Spirostomum teres*, somewhat smaller than *S. intermedium*, occasionally with zoochlorellae, compact elliptical macronucleus; in stagnant fresh water

g *Folliculina moebiusi*, peristome with zone of membranellae like a wing, with pseudochitinous casing; marine

h *Nyctotherus ovalis*, ventral view with a long oral cone, macronucleus fixed with fibrils, contractile vacuole at the back; in gut of the kitchen cockroach *Blatta orientalis*

i *Blepharisma lateritium*, flattened, with undulating membrane on the right rim of the peristome, cytoplasm pigmented reddish; in fresh water, particularly among plants

k *Caenomorpha medusula*, flattened, bell-shaped with terminal spine, with long cilia outside on the rim of the bell, spiral adoral zone of membranellae within; in standing fresh water

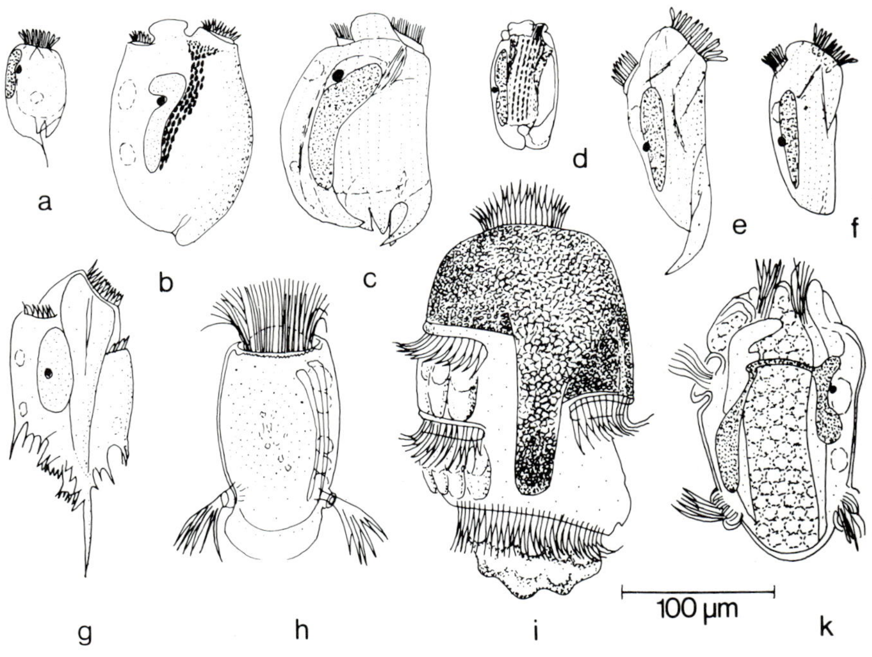
a
b
c
d
e
f
g
h
i
100 µm
k

igure 83. **Spirotricha, Entodiniomorpha** (Magnification 205 ×)
a *Entodinium caudatum*, clockwise anterior adoral zone, strong pellicle, with spine-like spurs at the back, in the rumen of cattle and sheep
b *Eudiplodinium maggii*, dorsal and adoral zone of cilia at the front (cirri and membranellae), a skeletal plate beside the macronucleus, 2 contractile vacuoles; in cattle, sheep and reindeer
c *Diplodinium dentatum*, dorsal and adoral zones of cilia at the front, pellicle toothed at the back, no skeletal plate; in rumen of cattle
d *Ostracodinium dentatum*, with broad skeletal plate, pellicle with broad caudal lobes; in cattle
e *Epidinium caudatum*, with dorsal and adoral zones of cilia. Skeleton of 3 plates, pellicle with terminal spine; in cattle, dromedary, reindeer and wapiti
f *Entodinium ecaudatum* like *E. caudatum* but without a terminal spine; in cattle, sheep and reindeer
g *Ophryoscolex quadricoronatus*, with adoral and extended dorsal zone, 3 meeting skeletal plates, posterior end jagged with terminal spine; in various sheep
h *Cycloposthium bipalmatum*, 2 further (posterior) zones of cirri in addition to the broad adoral zone, armour of 2 broad lateral plates with slender channel-like skeletal elements; in the caecum and colon of the horse
i *Troglodytella gorillae*, 4 further rows of cirri in addition to the adoral zone, armour basically as in *Cycloposthium*; in the wild, in the caecum and colon of the gorilla
k *Tripalmaria dogieli*, skeleton seen from the right, with marked thickenings, 2 dorsal and 1 ventral zone in addition to the adoral zone of cirri. In colon of the horse

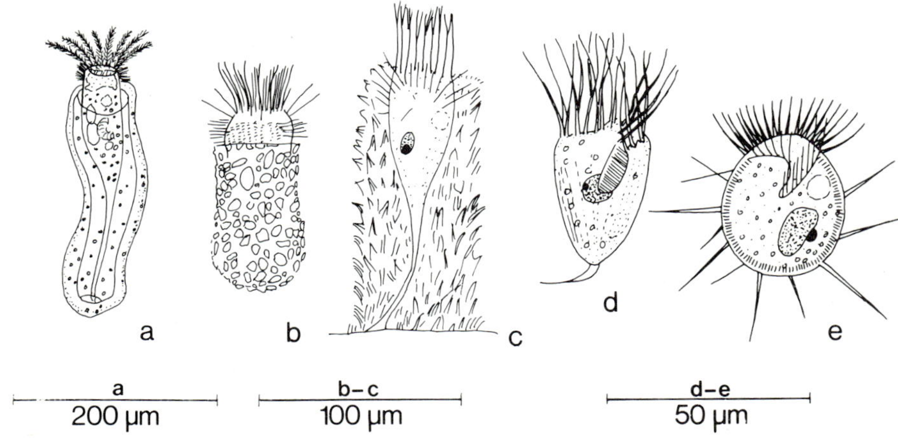

Figure 84. **Spirotricha, Oligotricha.** a Magnification 140 × , b–c Magnification 275 × , d–e Magnification 550 ×

a *Tintinnidium fluviatile*, with stalk-like contractile posterior end attached in a gelatinous coat, pinnate adoral zone of membranellae; at the surface of large fresh-water lakes, seldom attached

b *Codonella cratera*, capsule with foreign bodies; fresh-water species

c *Tintinnidium semiciliatum*, capsule encrusted, adoral zone of membranellae slit into 5–6 parts; in fresh water, at the surface of ponds

d *Strombidium calkinsi*, without capsule, peristome ventral, reaching as far as the middle of the cell, tapering processes at the front, thorn-like appendage at the back; in brackish and salt water

e *Halteria grandinella*, without capsule, spherical, long rigid bristles behind the adoral zone, rapid darting motion here and there then motionless again. In marshy fresh water

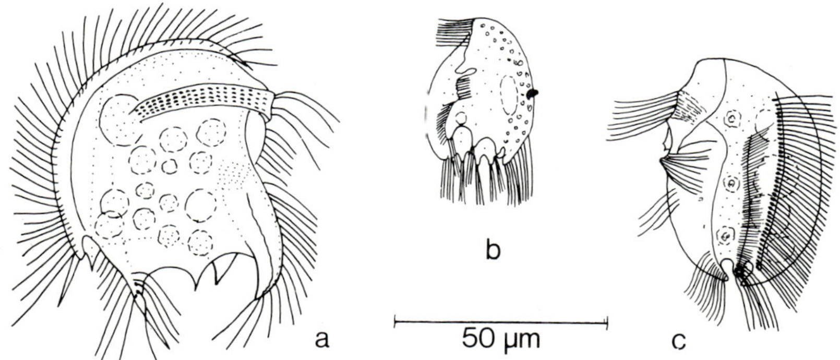

Figure 85. **Spirotricha, Odontostomata** (Magnification 550 ×)
 a *Saprodinium dentatum*, flattened with armour-like tough pellicle with anal teeth; in fresh-water putrescent mud
 b *Epalxis mirabilis*, small and flat, armour open at the back; in stagnant fresh water
 c *Pelodinium reniforme*, armour with 3 teeth at the back; in stagnant fresh water

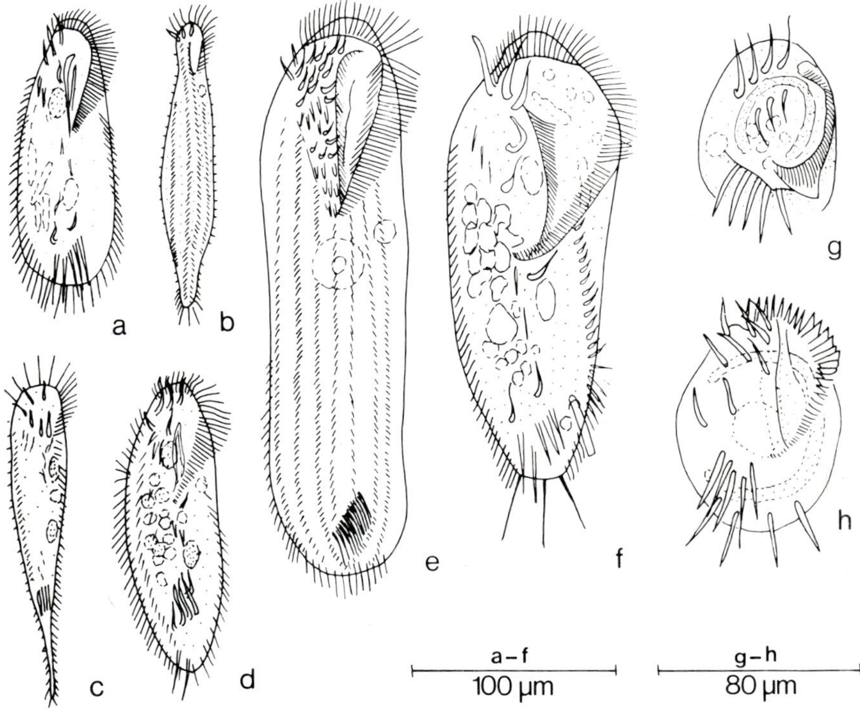

Figure 86. **Spirotricha, Hypotricha.**
a–f Magnification 205 × , g–h Magnification 345 ×
a *Oxytricha fallax*, anterior adoral zone, 8 frontal cirri, 5 ventral cirri and 5 anal cirri; in marsh water
b *Uroleptus limnetis*, cell tapering at the back, several ventral rows of cirri in addition to three frontal cirri; in fresh water among plants
c *Urosoma caudata*, like *Oxytricha* but the posterior end tapering into a tail; in standing fresh water
d *Pleurotricha lanceolata*, with 8 frontal cirri, 5 ventral and 5 anal cirri, with two additional ventral rows of cirri; in standing water among rotting leaves
e *Urostyla grandis*, with several frontal and anal cirri and numerous ventral rows of cirri, macronucleus of 100 or more separate pieces; in marshy water
f *Stylonychia mytilus*, ciliature of the ventral side as in *Oxytricha*, but 3 strong tail bristles; frequent in marsh water
g *Aspidisca lynceus*, ventral ciliate adoral zone, with a further 4 and 3 cirri at the front and 5 cirri at the back; in fresh water and sea
h *Euplotes patella*, 9–10 cirri at the front and 9–10 cirri at the back in addition to the adoral zone, including 4 marginal cirri at the back; in standing water among plants

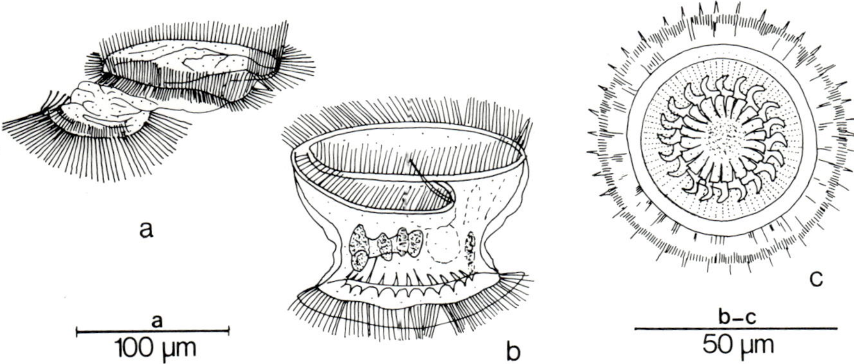

Figure 87. **Peritricha, Mobilia.**
a Magnification 205 × , b–c Magnification 550 ×
a *Urceolaria mitra*, upper adoral zone of cilia displaced in relation to the
lower adhesive disc; on fresh-water planaria
b *Trichodina pediculus*, adhesive ring above the adhesive disc developed as
a crown of hooks; on fresh-water polyps (*Hydra*) and tadpoles
c *Trichodina (Cyclochaeta) domerguei*, flatter than *T. pediculus*, figure shows
the underside of the adhesive disc with strong crown of hooks and adhesive
ring; on gills and skin of fresh-water fish, also in aquaria

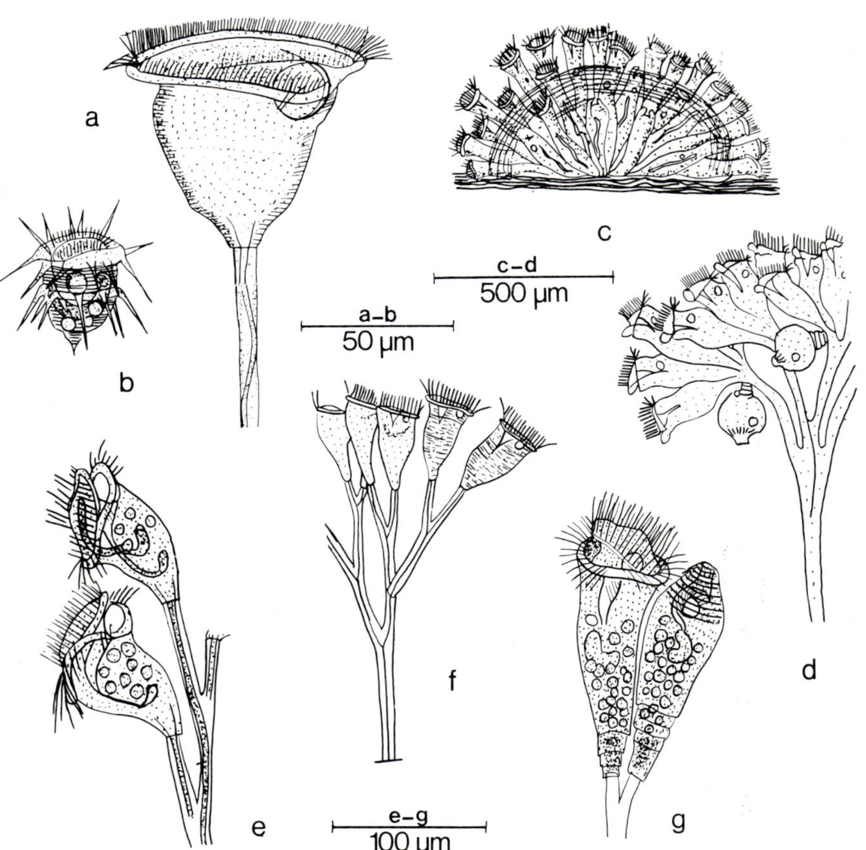
a
b
c
c-d
500 μm
a-b
50 μm
d
e
f
g
e-g
100 μm

Figure 88. **Peritricha, Sessilia (Aloricata)**
a–b Magnification 410 × , c–d Magnification 48 × , e–g Magnification 205 ×

a *Vorticella campanula*, living singly on a contractile stalk with stalk muscle, adoral zone surrounded by strong peristome swelling; in fresh water, on water plants.

b *Hastatella radians*, with crown of spines round the ciliated peristome rim and the middle of the cell, only occasionally attached by the posterior end; in standing fresh water

c *Ophrydium sessile*, with contractile cylindrical neck, as a colony in a common gelatinous coat (transition to **Loricata**, figure 89); on fresh-water plants

d *Opercularia plicatilis*, colony with branched stalk, peristome narrowed and without swelling; in standing waters

e *Carchesium polypinum*, colony with branched, very contractile stalk, in which the stalk muscles are not united together; on plants, stones, etc., in fresh water

f *Zoothamnium adamsi*, like *Carchesium* but the stalk muscles combined together (combined contraction); on aquatic plants (Cladophora).

g *Epistylis plicatilis*, branched colonies with noncontractile stalks, on plants and animals in fresh water.

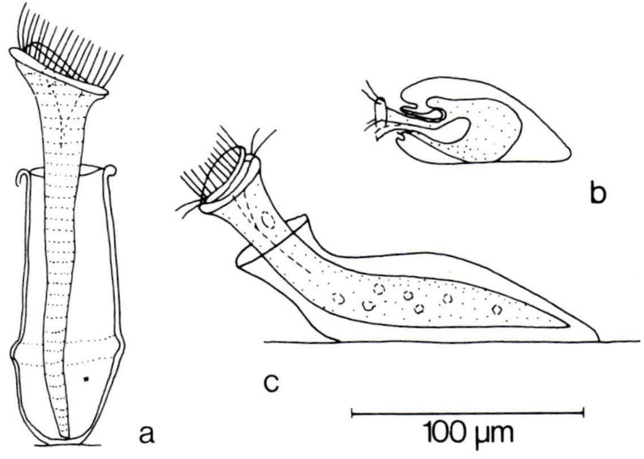

Figure 89. **Peritricha, Sessilia (Loricata)** (Magnification 275 ×)
a *Vaginicola annulata*, contractile in strong pseudochitinous capsule; in ponds
b *Lagenophrys vaginicola*, capsule with narrowed opening with a cover, to which the animal is attached by the peristome rim; on the crustaceans *Cyclops minutus* and *Canthocamptus* sp.
c *Platycola longicollis*, contractile in a laterally attached capsule with erect neck region; in fresh water

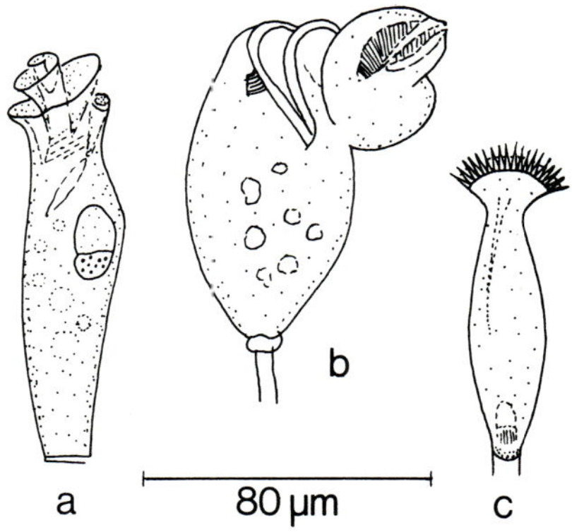

Figure 90. Chonotricha (Magnification 345 ×)
a *Spirochona gemmipara*, peristome with funnel-shaped ectoplasmic structure with ciliated base, otherwise aciliate; on the gills of *Gammarus pulex*, attached
b *Chilodochona quennerstedti*, peristome with lip structures (budding above right); in fresh water with long stalk on crustaceans (*Portunus depurator*, *Ebalia turnefacta*)
c *Heliochona scheuteni*, peristome funnel with needle-like border; in salt water on *Gammarus locusta*.

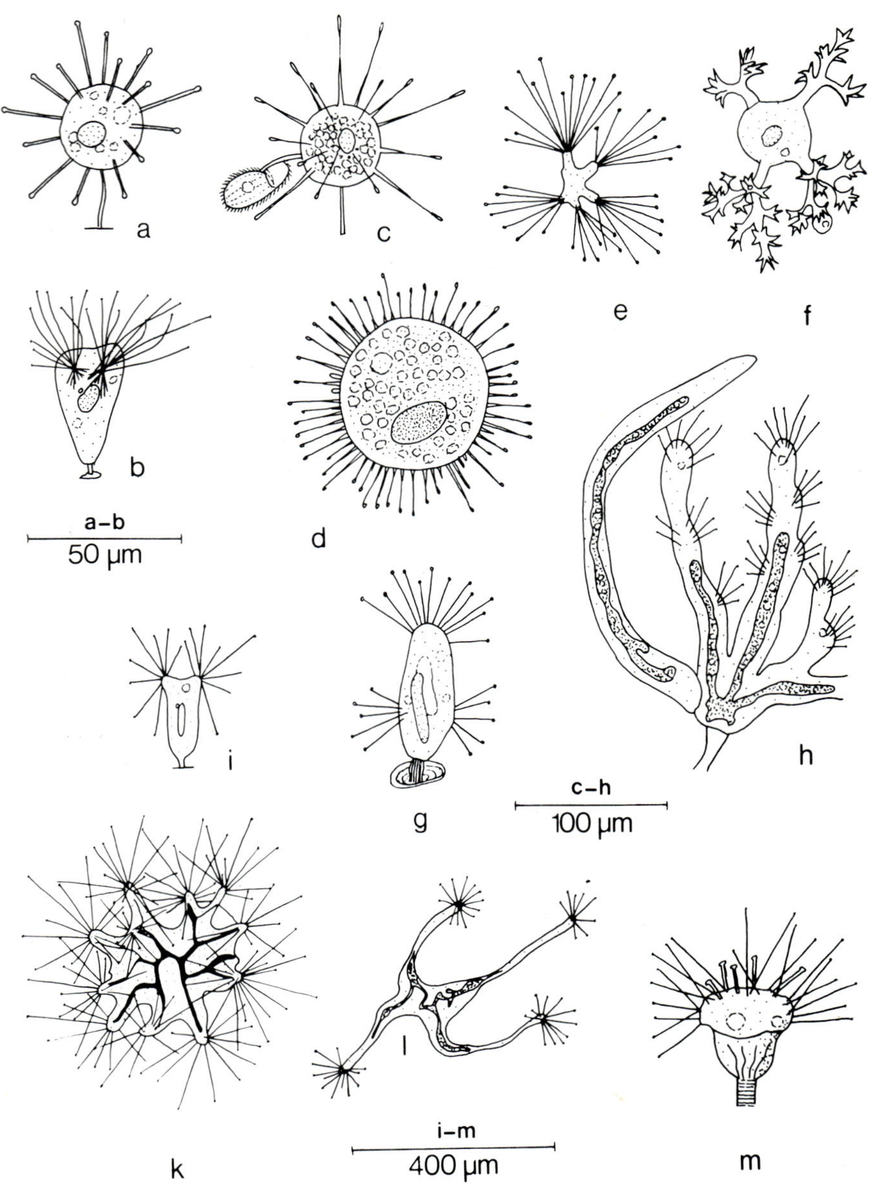
a
c
e
f
b
a–b
50 µm
d
i
g
c–h
100 µm
h
k
l
i–m
400 µm
m

Figure 91. **Suctoria**
a–b Magnification 410 × , c–h Magnification 205 × , i–m Magnification 70 ×
a *Podophrya fixa*, spherical, attached by a stalk, with uniform tentacles, division with swarmer formation; on algae in fresh water
b *Tokophrya infusionum*, narrowed at the bottom, attached, with 2 tufts of tentacles, reproduction by internal swarmer formation; common in infusions
c *Parapodophrya typha*, like *Podophrya* but division without swarmer formation (here sucking out the contents of a ciliate); in salt water
d *Sphaerophrya soliformis*, spherical, unstalked. Reproduction by binary fission with or without swarmer formation; in stagnant fresh water
e *Trichophrya epistylidis*, irregular outline, tufts of tentacles on bulges, without stalk. Reproduction by internal swarmer formation; in fresh water on the stalks of *Epistylis plicatilis* (figure 88 g) and aquatic plants
f *Dendrocometes paradoxus*, flat, contractile arms with spherical tentacles. Reproduction by internal budding. On the gills of *Gammarus pulex* and other **Gammaridae**
g *Discophrya elongata*, with rigid tentacles, reproduction by internal budding; in fresh water on the shells of *Paldina vivipara*
h *Dendrosomides paguri*, with vermiform arms, reproduction by cutting off individual arms (external budding); in salt water on the crabs *Eupagurus excavatus* and *E. cuanensis*.
i *Acineta lacustris*, attached with capsule, tentacles in two tufts; in ponds
k *Lernaeophrya capitata*, tentacles on finger-like processes which rise up around the margin, large branched macronucleus, reproduction by internal budding; in fresh water, also in brackish water on the hydrozoan *Cordylophora lacustris*
l *Dendrosoma radians*, with long simple or branched arms with tentacles, in a mucilagenous coat, reproduction by internal budding; in fresh water on plants
m *Ephelota gemmipara*, head with long pointed ensnaring tentacles and shorter sucking tentacles, on long stalk, reproduction by external budding; marine on hydroids, Bryozoa and algae

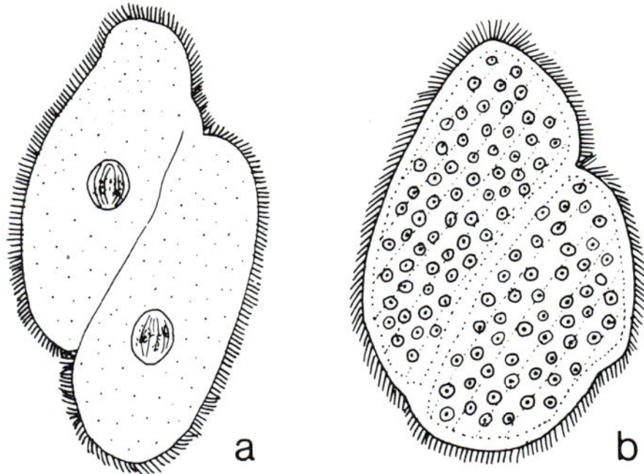

Figure 92. **Reproduction of the Protociliates:** plasmotomy in the form of diagonal binary fission
   a *Zelleriella elliptica*
   b *Opalina ranarum*

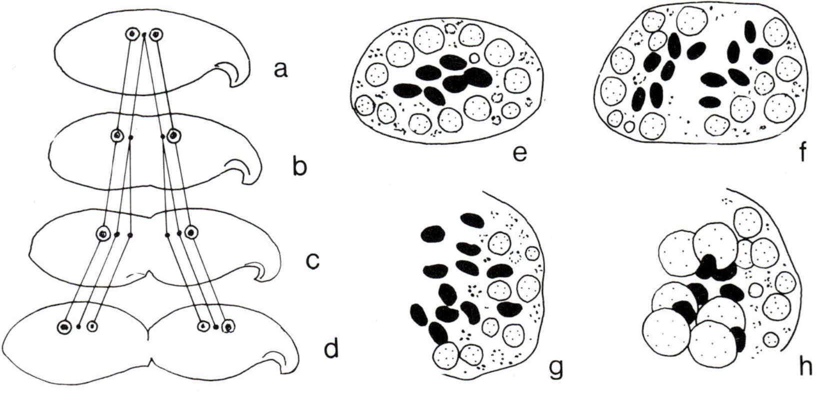

Figure 93. **Reproduction of the Euciliates with a diploid macronucleus**

a–d *Loxodes rostrum*
   a normal vegetative form with 1 micronucleus and 2 macronuclei with nucleolus
b–d division with regeneration of the nuclei (formation anew of the missing macronucleus from a micronucleus)
e–h *Tracheloraphis phoenicopterus*
   e assembled resting nucleus with 6 (black) micronuclei and 12 macronuclei
   f onset of nuclear division after doubling of the micronuclei
   g doubling of the micronuclei in one half of the nucleus
   h change of 6 micronuclei into macronuclei.

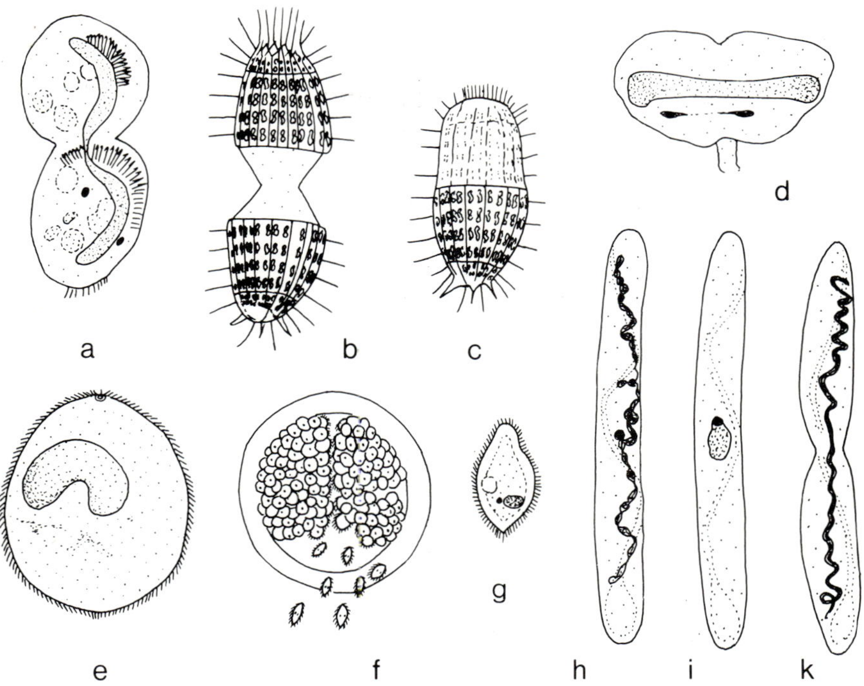

Figure 94. **Reproduction of the Euciliates with a polyploid macro-nucleus**

 a transverse division of *Euplotes charon*, the macronucleus being cut through

b–c division of *Coleps hirtus*

 b transverse division

 c completion of the missing half

 d longitudinal division of *Carchesium polypinum*

e–g reproduction of *Ichthyophthirius multifiliis* **(Holotricha, Hymeno-stomata)**

 e vegetative form in the skin of an infected fish (500–800 µm)

 f Reproduction cyst with 2 daughter cells, in which numerous swarmers arise by further divisions

 g free ciliate swarmer

h–k division of *Spirostomum ambiguum*

 h normal vegetative stage with chain-like macronucleus

 i macronucleus forming into a ball

 k transverse division of the cell. Macronucleus, once more elongate, being cut through

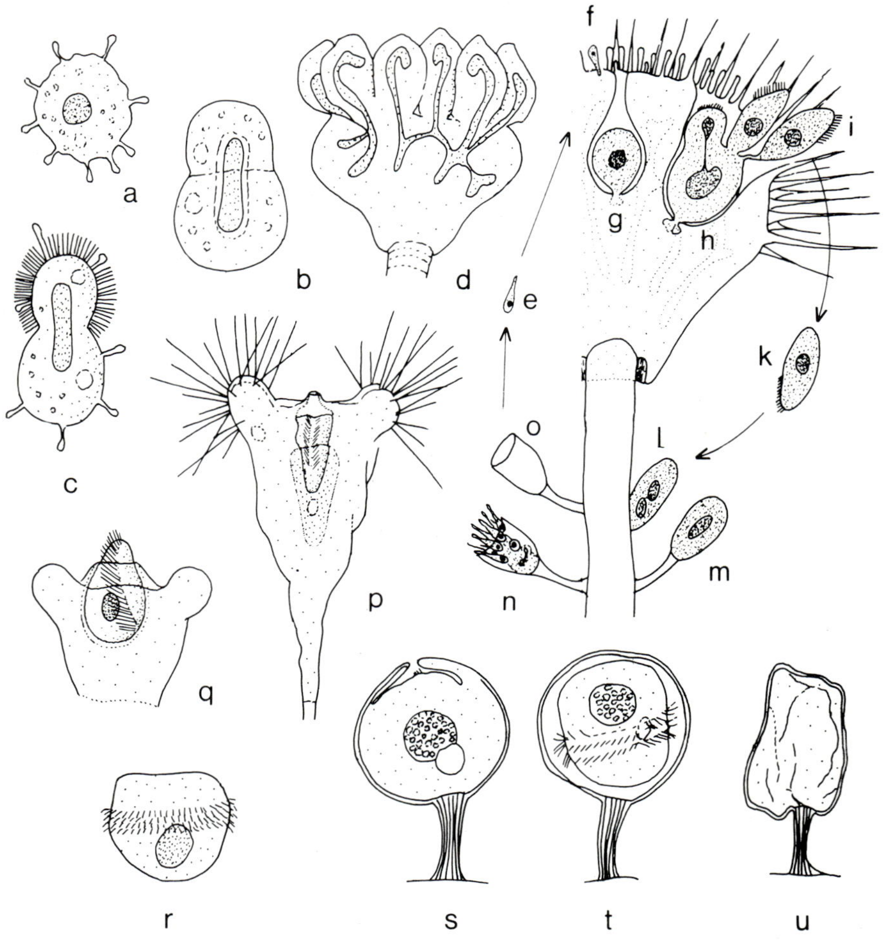

Figure 95. **Reproduction of the Suctoria.**
a–c *Sphaerophrya pusilla*
  a normal vegetative form with tentacles
  b binary fission
  c division with development of a ciliate bud
  d *Ephelota gemmipara*, multiple external budding
e–o *Tachyblaston ephelotensis*, parasite on *Ephelota gemmipara* with alternation
of generations
  e free dactylozoite with tentacle
  f penetration of the dactylozoite into the head of *E. gemmipara*
  g developed parasite after penetration
  h–i external budding with swarmer production
  k free ciliate swarmer
  l attachment of the swarmer on the stalk of *E. gemmipara*
  m developed free-living stage with capsule
  n repeated division with development of dactylozoites with tentacle
  o empty capsule
p–r *Tokophrya quadripartita*
  p development of an internal bud in a brood chamber
  q release of ciliated bud (swarmer)
  r free swarmer
s–u *Tokophrya cyclopum*, total conversion into a swarmer bud
  s onset of formation of brood chamber
  t mature swarmer
  u relinquished, empty coat

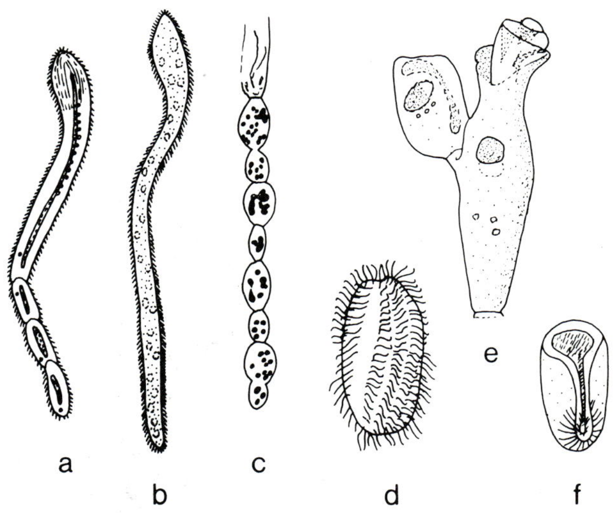

Figure 96. **Reproduction of Euciliates by budding**
  a *Radiophrya hoplites* (Holotricha, Astomata) chain-like budding at the posterior end
b–d *Chromidina elegans* (Holotricha, Apostomea)
  b trophont before budding
  c tomont, chain-like production of buds at the posterior end of the trophont
  d released ciliate bud: tomite
e–f *Spirochona gemmipara*
  e production of buds laterally on the mother animal
  f released bud, with foot disc below

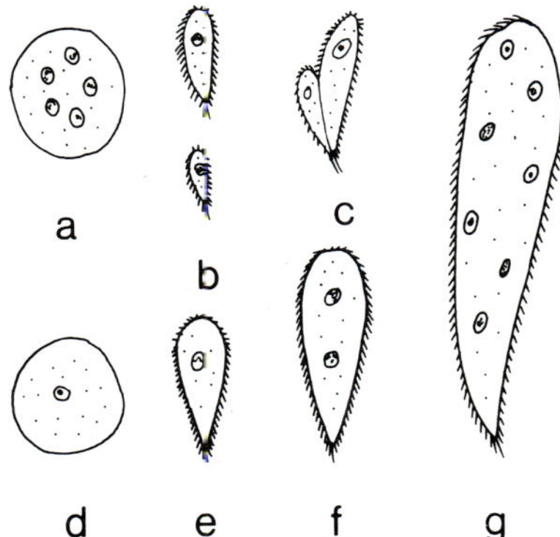

Figure 97. **Sexuality in Opalina ranarum (Protociliata)**
  a multinucleate cyst
  b macrogamete and microgamete
  c fusion of the gametes
  d cystozygote
  e young vegetative form
  f–g maturation with multiplication of nuclei

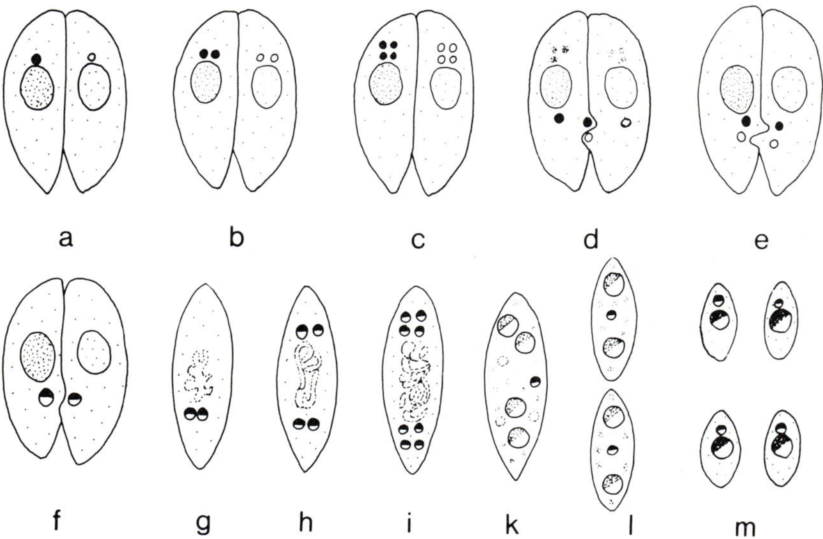

Figure 98. **Sexuality in the Euciliates.** Conjugation of *Paramecium caudatum*.

a pairing of the conjugants each with 1 micronucleus and 1 macronucleus
b first pregamic division of the diploid micronucleus (equatorial division)
c second pregamic division of the micronucleus (meiosis)
d disintegration of 3 micronuclei in each conjugant, postmeiotic division of the remaining fourth haploid micronucleus into stationary nucleus and migratory nucleus
e each migratory nucleus is transferred into the other conjugation partner
f fusion to a diploid zygotic nucleus (synkaryon) in each
g first postzygotic division of the diploid micronucleus in the exconjugant now no longer paired, disintegration of the old macronucleus
h second postzygotic division of the micronucleus in the exconjugant
i third postzygotic nuclear division
k growth of 4 micronuclei into macronuclei, 3 micronuclei disintegrated
l distribution of the macronuclei between 2 daughter animals after previous division of the remaining micronucleus
m restoration of normal nuclear conditions by further binary fission after previous division of the micronucleus

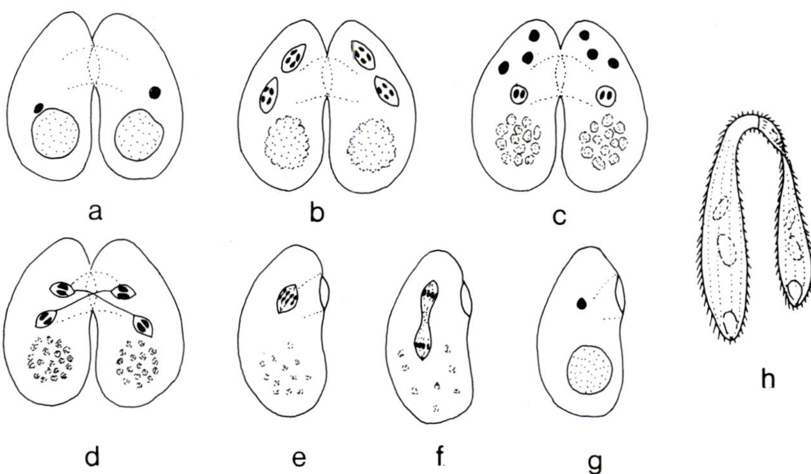

Figure 99. **Sexuality in the Euciliates**
a–g conjugation of *Chilodonella uncinata*
   a association of the conjugants
   b situation after the first pregamic division of the micronucleus (equatorial division), onset of the disintegration of the macronucleus
   c after the second pregamic division of the micronucleus (meiosis), disintegration of 3 micronuclei and the macronucleus
   d after the third pregamic division of the now haploid micronucleus, each conjugant now having 1 migratory nucleus and 1 stationary nucleus
   e diploid zygotic nucleus (synkaryon) in an exconjugant
   f postzygotic division of the micronucleus
   g normal nuclear complement after one micronucleus has changed to a macronucleus
   h *Lacrymaria apiculatum*, anisogamous conjugation.

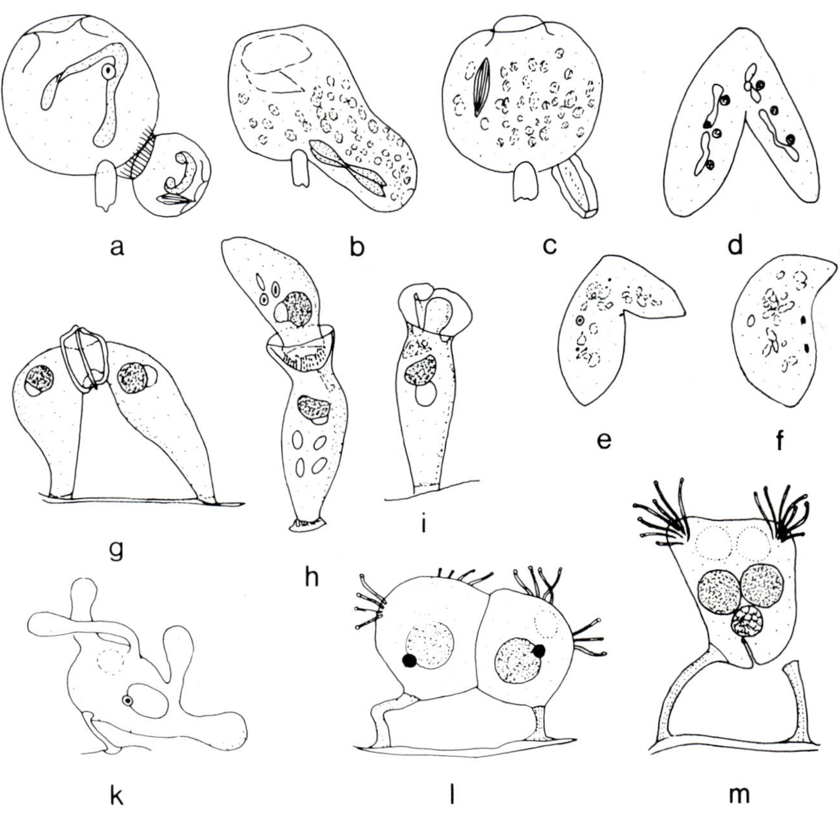

a
b
c
d
e
f
g
h
i
k
l
m

Figure 100. **Sexuality in the Euciliates**
a–c anisogamous conjugation in *Vorticella nebulifera*
    a attachment of the free microconjugant on to the macroconjugant which
is attached to the substratum
    b fusion of the conjugants and production of the stationary and migratory
nuclei
    c macroconjugant with synkaryon and adjacent degenerated micro-
conjugant
d–f total conjugation in *Urostyla polymicronucleata,*
    d association of the conjugants
    e fusion of the conjugants
    f completion of the total fusion
g–i total conjugation in *Spirochona gemmipara*
    g the conjugants making contact
    h onset of fusion after one of the conjugants has become detached
    i remaining exconjugant after total fusion
k–m conjugation in the suctorian *Tokophrya cyclopum*
    k development of conjugation bridges to make contact with a conjugation
.partner
    l two neighbouring conjugants after contact has been established
    m detachment of one conjugant before total fusion

THE CELL ORGANELLES OF THE PROTOZOA

The Cell Organelles of the Protozoa

The nucleus, its somatic function, and the endoplasmic reticulum

The importance of the nucleus in the living functions of the cell becomes apparent when it is removed from the cell. This can be achieved either by sucking out the nucleus with a micropipette in a micromanipulator or, since the Protozoa are generally able to regenerate, more simply through merotomy, cutting the cell into nucleated and anucleated pieces. After the cell has been divided up in this way (figure 101) the wound closes in all cases. Whilst motility, function of the contractile vacuole and, in some species, at first also food uptake are retained, the anucleate forms always lack the capacity to digest. Comparative studies with the ciliate *Stentor* have shown that anucleate forms die for this reason in the same time as it takes nucleated forms to die of starvation when deprived of a food source. If amoebae which have accumulated glycogen and fats in the cytoplasm before merotomy are maintained after the operation in a medium lacking nutrients, the reserves in the nucleate half are largely broken down within a week, whilst the anucleate amoeba can no longer utilize the reserves.

Morphological regeneration is exemplified by *Amoeba proteus* and *Stentor coeruleus* in figure 101. Only the amoeba with a nucleus remains functioning (figure 101 b, d). In a ciliate, all fragments achieve complete regeneration if they contain parts of the segmented macronucleus (e–h); even single links of the chain-like nucleus are sufficient. In contrast, the segments lacking a nucleus are capable of surviving for only a short while (i–n). In *Stentor coeruleus* the presence of the micronucleus in a

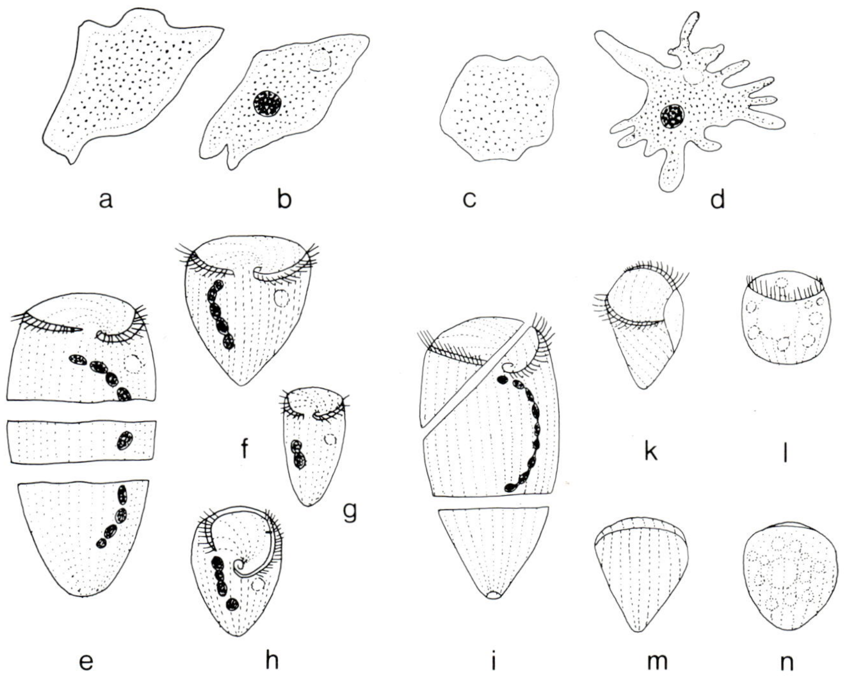

Figure 101. **Merotomy experiments**
a–d *Amoeba proteus* (*Chaos diffluens*) cut up into a nucleate and an anucleate piece
    a anucleate half
    b nucleate half
    c regeneration of the anucleate half after 2 days without food uptake
    d complete regeneration of the nucleate half with pseudopodia, food uptake and capacity for division
e–n *Stentor coeruleus* cut into pieces
    e dissection into three parts with differing numbers of macronuclei
    f–h completed *Stentor* cells with full regeneration
    i Dissection into 1 nucleate and 2 anucleate pieces. Whilst the middle, nucleate, piece completely regenerates as in f–h, the anucleate sections show only wound closure followed by vesicular degeneration (k–n).

merotomy fragment has no influence on regeneration, whilst in the ciliate *Euplotes patella* only pieces which also have a micronucleus are capable of life in the long term. Even so, fragments of ciliates with only the micronucleus cannot regenerate. These studies show the significance of the cell nucleus in the somatic

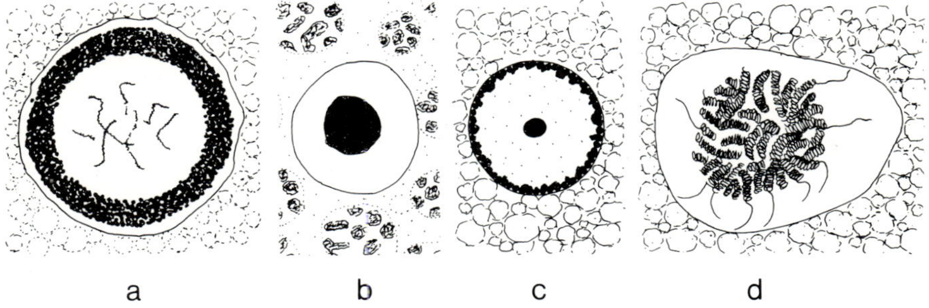

Figure 102. **Nuclear structure of Protozoa during interphase** (resting nucleus) after haematoxylin staining

    a *Myxotheca arenilega* (Foraminifera). A strongly developed ring of nucleolar substance in the basic nuclear substance (karyoplasm) within the nuclear boundary (nuclear membrane, nuclear coat), and distinct chromosomes in the middle

    b *Limax amoeba* (Amoebina). Within the nuclear membrane all the chromatin (chromosome and nucleolar substance) lies compacted together in the centre of the karyoplasm (vesicular nucleus)

    c *Entamoeba histolytica* (Amoebina). Chromatin closely applied to the nuclear membrane in a ring. and also centrally as an inner body in the karyoplasm. Inner body consists only of nucleolar substance. Chromosomal substance seen only in outer chromatin ring

    d *Pseudotrichonympha intraflexibilis* (Polymastigina). Spiral structure of chromosomes is clear.

functions of the cell; these functions are dependent upon the macronucleus in the ciliates, which is therefore called the somatic nucleus.

An analysis of nuclear structure can be made with the aid of light microscopy or electron microscopy. Because of its greater optical density, the nucleus of many Protozoa can be distinguished from the cytoplasm by light microscopy, even in unstained preparations. For example, the nucleus of many entamoeba species appears as a bright shining ring in transmitted light, whilst, under phase contrast, it appears darker than the cytoplasm. A clearer picture of the nuclear structure can be obtained by appropriate staining after fixing. By this means, in suitable cases, four basic elements may be discerned in the nuclear structure, as can be seen, for example, in the gamont nucleus of the foraminiferan *Myxotheca arenilega* in figure 102 a. The nuclear membrane separates the nucleus from the surrounding cytoplasm, and the nucleolar substance forms a substantial peripheral ring, here stained with iron haematoxylin. Within the nucleus lie the stained chromosomes in various shapes and sizes, and all lighter spaces in between are filled with nucleoplasm. In the nuclei of most Protozoa the chromosomes can be discerned only when the nucleus enters mitotic division. In the intervening period, in the resting or interphase nucleus, the structural elements of the chromosomes are so finely distributed that they do not show clearly.

The arrangement of the basic elements in the nucleus can differ widely. Figure 102 b shows a very simple nuclear structure, the vesicular nucleus of a limax amoeba, in which the "chromatin" component of the nucleus (which can be stained with haematoxylin or other basic stains) is a compact structure in the centre of the nucleus, while a broad zone of nuclear sap or nucleoplasm lies peripherally. In the nucleus of *Entamoeba* species, on the other hand, the chromatin is divided between small central internal bodies and a peripheral chromatin ring under the nuclear membrane. The particular staining properties of the chromosomes and nucleolar substance within the nucleus derives from their nucleic acid content. Desoxyribonucleic acid (DNA) is the major component of the chromosomes, which contain only a small amount of ribonucleic acid (RNA), while the ability of the nucleolar substance to take up the customary nuclear stains is a function of the RNA content.

Differentiation between the DNA- and RNA-containing components of the nucleus is best achieved by the use of the Feulgen reaction. The purines guanine and adenine separate out from the DNA, with the formation of apurinic acids, when the DNA is hydrolyzed in warm dilute hydrochloric acid. The free aldehyde groups react with fuchsin-sulphurous acid, which is added as Schiff's reagent, with the production of a red-violet pigmentation. Since this reaction occurs only with DNA, the RNA components of the nucleus remain unstained. Hence the chromosomal substance is Feulgen-positive and the nucleolar substance is Feulgen-negative. The Feulgen reaction would therefore stain the central chromosomes violet in the *Myxotheca* nucleus (figure 102 a) while the outer chromatin ring, the nucleolus, would remain unpigmented. In contrast, when Feulgen staining is applied to the *Entamoeba* nucleus (figure 102 c) the chromosomal substance is demonstrable only in the

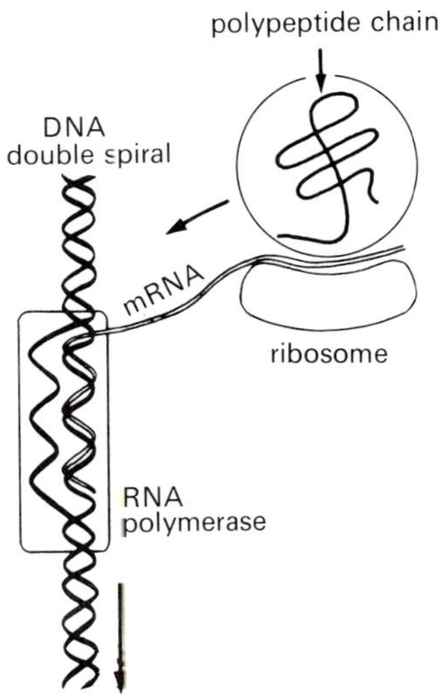

Scheme 1: Diagram illustrating transcription during protein synthesis (from Lehninger).

outer chromatin ring, while the central inner body of the nucleus proves to be a nucleolus. Differentiation between DNA and RNA can also be achieved by methyl green-pyronin staining (Unna and Pappenheim). Since methyl green has a greater affinity for DNA, this turns green, whilst the RNA is stained red by the pyronin.

Despite the ease with which chromosomal material can be demonstrated by the Feulgen reaction, it is by no means always possible in the interphase nucleus. In the *Entamoeba* nucleus, the chromosomes are first seen within the nucleus during mitosis. Only a few Protozoa show their chromosomal arrangement in the resting nucleus. Apart from *Myxotheca arenilega*, the chromosomes of the Euglenoidina, Dinoflagellata, and of some Polymastigina (figure 102 d) can be seen in the resting nucleus, sometimes even in unstained preparations. Chromosomes represent, through the organized arrangement of the various genes, the hereditary factors which work together to determine the morphological and physiological specificity of particular species. The whole set of chromosomes is known as the **genome**. A nucleus in which each hereditary factor is

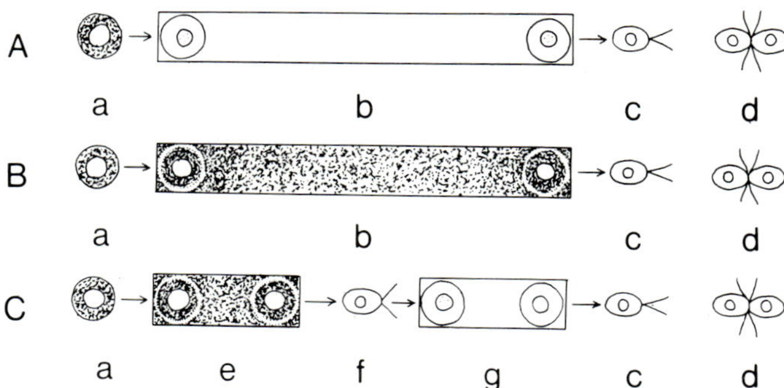

Figure 103. Diagram of the **alternation between haploid and diploid phases** after Grell. Black: diploid, white: haploid.
A: haplont with zygotic meiosis
B: diplont with gametic meiosis
C: heterophasic alternation of generations with intermediate meiosis
   a zygote
   b vegetative phase
   c gamete
   d fusion of two gametes
   e vegetative phase of the agamont
   f agamete
   g vegetative phase of the gamont

present only once, and which thus contains only one genome, is a **haploid** nucleus. Nuclei with 2 genomes are **diploid**; those with numerous genomes are **polyploid**. The genes present in the genomes, which are made up of DNA, determine all that happens in the cell. They supply the information for the synthesis of proteins which, as structural elements, participate in the construction of the cell and, as enzymes, make possible the many different metabolic processes of the cell. mRNA (messenger RNA) carries the information from the DNA to the ribosomes, the site of protein synthesis. Scheme 1 on page 187 shows this relationship. The upper arrow shows the direction of movement of the ribosomes.

The total number of chromosomes in a genome is very varied, so that great differences can occur within a single Order. For example in the Polymastigina there are species known to have 2, 8, 10, 12, 14, 24, 26, 48 and 60 chromosomes in the genome.

Most nuclei in the Protozoa contain 1 or 2 genomes and are thus haploid or diploid. During sexual processes all gametic nuclei are haploid (figure 103 c) and, in the ensuing karyogamy, their fusion (figure 103 d) produces zygotic cells which are diploid (a). If division of the zygote proceeds as a reduction division, so that a **zygotic meiosis** occurs (figure 103 A), the vegetative phases arising from it are all haploid. This behaviour has been seen in the Phytomonadina, some Polymastigina and the Sporozoa. If the zygote divides by a simple mitosis without chromosome reduction, all the following vegetative phases remain diploid (figure 103 B). This applies to the heliozoan genera *Actinosphaerium* and *Actinophrys*, and also to some Polymastigina and to all the micronuclei in the Euciliata. Nuclear division does not become meiotic until gametic nuclei are again produced, i.e. a **gametic meiosis** occurs. **Intermediate meiosis** in the Protozoa occurs only in the Foraminifera (figure 103 C). The zygote gives rise to the diploid vegetative phase of the agamont from which the young haploid agametes arise by meiosis; as haploid gamonts these produce the gametic cells. The vegetative phases of diploid agamonts and haploid gamonts thus form a heterophasic alternation of generations.

In addition to haploid and diploid nuclei, polyploid nuclei also occur in the Protozoa. Feulgen staining of the macronuclei of most euciliates shows a great abundance of DNA with a number of nucleolar zones lying in between. In diffuse macronuclei, e.g. as in *Nassula ornata* (figure 104 a), division configurations of the chromosomal structures can be seen in the macronucleus too, at the time of division of the micronucleus. These configurations indicating chromosomal division can also be seen in the slowly growing macronucleus of the suctorian *Ephelota gemmipara* and during macronucleus formation in exconjugants of *Bursaria truncatella,* and in *Stentor polymorphus.* The mass of chromosomal structures demonstrable here far exceeds the chromosomal content of the relevant diploid micronucleus from which the macronucleus develops anew after conjugation. In the production of the macronucleus there must therefore be repeated mitoses in the nuclear body, without nuclear division. This process is called **endomitosis** and it leads to endomitotic polyploidy. The nuclei produced in this way are also often called polyenergid nuclei. It is interesting to note that, as merotomy experiments show, the somatic functions of the highly developed euciliates are controlled by these polyploid nuclei and not by the diploid micronucleus. The relative propor-

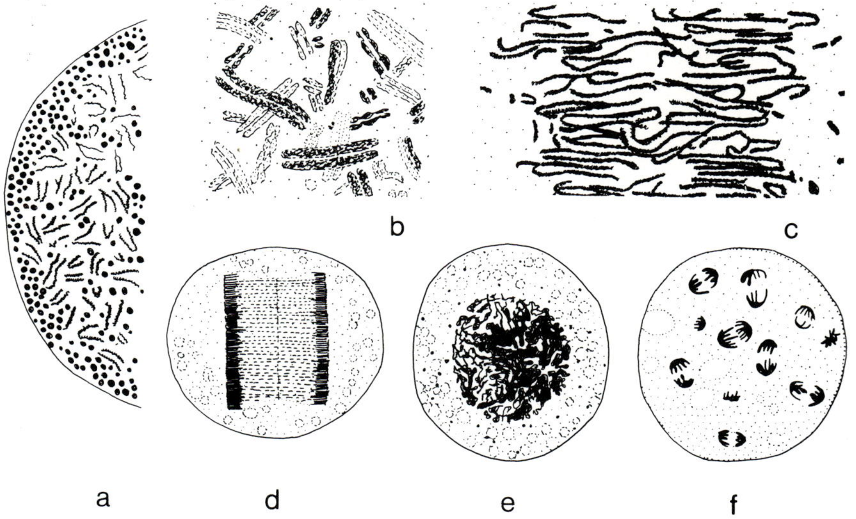

Figure 104. **Polyploid nuclei**
  a *Nassula ornata* (ciliate, Hypostomata). Picture of part of the macronucleus with chromosomes in division (endomitosis)
b–f *Aulacantha scolymantha* (Radiolaria)
  b Section of the nucleus in the central capsule with chromosomes in endomitosis
c–d binary fission of the nucleus
  c onset of binary fission, showing part of the equatorial plate with numerous genomes (high magnification)
  d drawing apart of the daughter plates
e–f nuclear production of the swarmer
  e dissolution of the nuclear membrane in the central capsule with the onset of decomposition of the primary nucleus for the formation of several small secondary nuclei
  f mitosis of the secondary nuclei formed, with the production of numerous swarmer nuclei.

tion of DNA in the cell is obviously not without significance with respect to the metabolic processes. Spectrophotometric measurements show, by comparison with the micronucleus which is diploid at the time, that the level of polyploidy varies widely among different species. While *Stylonychia mytilus* (figure 86 f) is only 64-ploid, *Nassula ornata* (figure 104 a) is 230-ploid, *Paramecium aurelia* 860-ploid, *Bursaria truncatella* (figure 82 b) 5000-ploid and *Ichthyophthirius multifiliis* (figure 94 e) even 12,600-ploid. It is not only the quantity of DNA but also the area of surface contact between the nucleus and the cytoplasm that is raised in many cases by the form of the macronucleus. Few ciliates have a round macronucleus, which has the smallest surface area in relation to its volume (figures 78 b, 79 a, c). In many cases the macronucleus has an oval shape (figures 75 c, 76 a, 77 c, 83 g), and further enlargement of the nuclear surface follows elongation of the nucleus (figures 80 a–d, 83 a–f), branching (figures 91 h, k, l, 93 d), chain formation (figures 82 a, d, e, 101 e, i), or the production of several individual macronuclei (figures 75 f, g, 93 a–d).

Polyploid nuclei occur not only in the ciliates but also in the Tripylea of the Radiolaria. A highly magnified section of the nucleus from the central capsule of *Aulacantha scolymantha* (figure 104 b) shows the splitting of chromosomes during endomitotic polyploidy. If the central capsule divides during binary fission (figure 43 e), an equatorial plate bearing a large number of ribbon-shaped chromosomes occurs at nuclear division (figure 104 c); these chromosomes are distributed among the daughter nuclei with the formation of daughter plates (figure 104 d). For swarmer production in the Radiolaria (figure 43 f, g) the membrane of the primary nucleus first disintegrates in the central capsule. After this, the numerous centrally-situated chromosomal genomes are distributed throughout the plasma of the central capsule (figure 104 e) and become secondary nuclei which undergo further mitoses (figure 104 f) to produce numerous swarmer nuclei. In many cases an increase in the chromosomal complement is achieved in the homokaryotic Protozoa only by them becoming multinucleate. Examples of this are provided by the Distomatidae (figure 18 g, h) of the Protomonadina, by the Polymastigina (figure 20 d, e), and by the amoeba *Pelomyxa palustris* (figure 27 a), the heliozoan species *Actinosphaerium eichhorni* (figure 32 a) and the homokaryotic ciliates (figure 73).

Although most euciliates have polyploid macronuclei, some holotrichous genera have only diploid macronuclei like the heterokaryotic Foraminifera. This condition has already been mentioned in dealing with the reproduction and sexuality of the ciliates, and in these cases the macronuclei are not able to divide; they are always formed anew from diploid micronuclei (figure 93). In these ciliates, too, an increase in DNA and in the

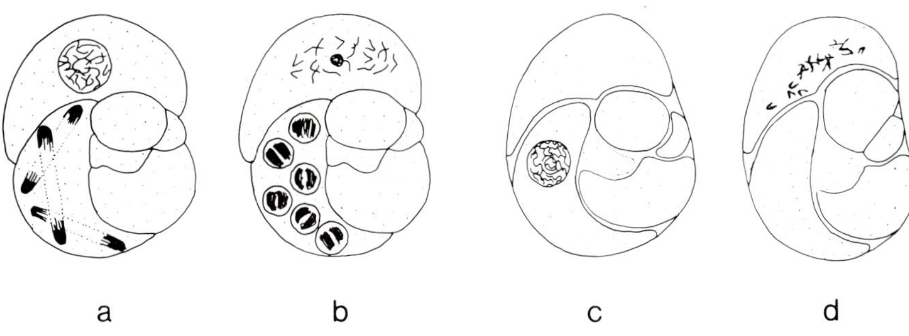

Figure 105. **Diploid macronuclei in Foraminifera**
a–d *Rotaliella roscoffensis*
   a meiotic division of the 3 micronuclei (anaphase with developed agamonts, clear chromosomal structure in the diploid macronucleus in the outer chamber
   b disintegration of the macronucleus, mitosis of micronuclei in metaphase
   c anomalous agamont without micronuclei, only with a macronucleus
   d disintegration of the macronucleus, when the agamont has reached the size appropriate for meiosis

surface area of the nucleus is achieved by the multinucleate condition (figures 74 f, 75 f).

In the heterokaryotic Foraminifera the chromosomes present in the macronucleus are clearly visible during meiosis of the micronuclei (figure 103 C: transition from e to f). When the micronuclei enter meiotic division, the macronucleus of the agamont in *Rotaliella roscoffensis* breaks up and disintegrates, thereby releasing the chromosomes (figure 105 a, b). Aberrant agamonts which have a macronucleus but no micronucleus can occasionally arise (figure 105 c). Like the ciliates, they are able to survive and grow without a micronucleus; but, when the agamont has reached the size at which meiosis normally begins, the macronucleus disintegrates just as it would in the course of meiosis (figure 105 d), and the resultant cessation of somatic function leads to the death of the cell. Again as with ciliates, Foraminifera which have only micronuclei can never develop. In contrast to these Foraminifera, ciliates with a polyploid macronucleus have an evidently unlimited capacity to live and reproduce. The capacity of the polyploid macronucleus for division has already been shown in figure 94. Thus the ciliate races lacking a micronucleus, which are occasionally found in nature, can maintain themselves by asexual reproduction, as has been shown in cultures of *Tetrahymena pyriformis, Oxytricha hymenostoma, Oxytricha fallax, Urostyla grandis, Didinium nasutum, Colpoda steini* and *Paramecium bursaria*. These findings support the results of merotomy experiments, according to which the macronucleus is responsible for the somatic cell function. As already described, the Feulgen reaction serves to demonstrate the DNA constituent in a light-microscopic study of the cell. Bearing in mind that the genes of the chromosomes also control the somatic functions of the cell, it is interesting to know whether the positive demonstration of DNA in the protozoan cell can always be traced back to the chromosomes. That this is not invariably so is shown by the structural analysis of mitochondria, plastids and especially the kinetoplast of the Trypanosomatidae (figures 17 b–d, 18 d). This autonomous organelle lying near the basal body of the flagellum contains DNA in its core (figure 112), although chromosomal structures as found in the nucleus do not occur.

In contrast to the chromosomes, the nucleolar substance contains RNA but no DNA. The quantity of RNA in the nucleus is largely dependent on its physiological condition. Figure 106 a shows the nucleus of *Amoeba proteus* (*Chaos diffluens*) after staining with gallocyanin-chrome alum, which stains DNA and RNA. If the RNA is removed before staining, by means of the enzyme ribonuclease, the appearance is as in figure 106 b. Figure 106 c shows the appearance of the nucleus of the amoeba after mitosis where the RNA fraction has been used up during nuclear division. *Gregarina cuneata* (figure 49 b), a species found in the mealworm, develops to a considerable size during the growth period, and staining with the normal basic nuclear stains shows the presence of a very large nucleus (figure 106 d). When mitotic divisions begin after syzygy (figure 48 o), the RNA fraction of the nucleus disappears (figure 106 e). The DNA fraction, the "generative nucleus", which in this species was already visible, now enters into mitosis.

Merotomy experiments suggest that the RNA fraction of the nucleus is transferred to the cytoplasm. The ability of the cytoplasm to take up basic stains is reduced within a matter of days in segments lacking a nucleus; the RNA content of the cytoplasm has already fallen to 60% of normal after 10 days. If amoebae are fed on foodstuffs containing radioactive ^{32}P,

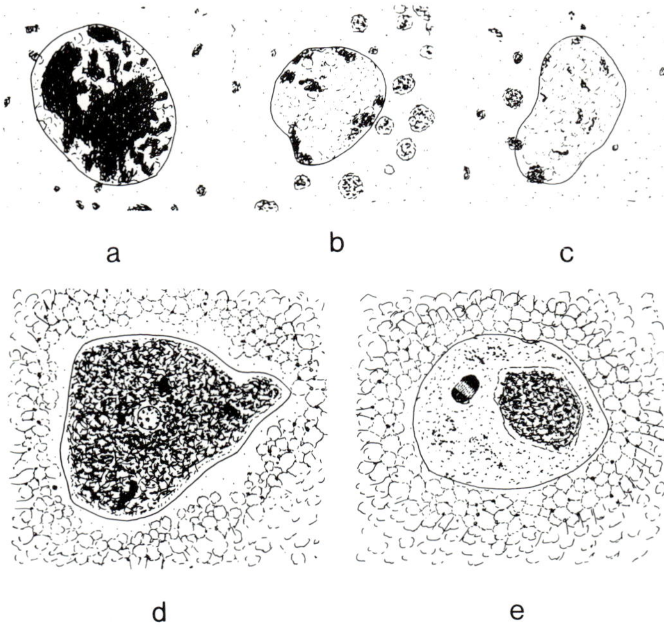

Figure 106. **Division of the chromosomal and nucleolar substance in the nucleus.**
a–c *Amoeba proteus* (*Chaos diffluens*), staining of the nucleus with gallocyanin-chrome alum
  a normal resting nucleus, staining after a 4-hour pretreatment solely with water
  b pigmentation after 4-hour pretreatment with ribonuclease
  c nucleus after mitosis, without pretreatment
d–e *Gregarina cuneata*
  d stained nucleus during the growth period of *Gregarina*; chromosomal substance ("generative constituent of the nucleus") inside, copious nucleolar substance outside
  e nucleolar substance disbanding as mitosis begins in the generative constituent which is now free

radioactivity is demonstrable in the nucleus. If such a nucleus is implanted into an untreated amoeba in a normal culture solution, the radioactivity is transferred to the cytoplasm. Further evidence for the importance of the nucleus in relation to the occurrence of RNA in the cytoplasm has been provided by experiments with ciliate *Tetrahymena pyriformis* (Hymeno-

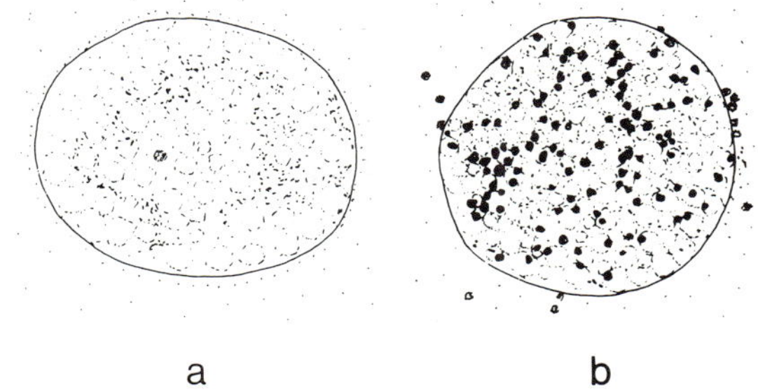

Figure 107. **Dependence of RNA production in the cytoplasm on the cell nucleus**
a–b *Tetrahymena* (ciliate, see figure 78 d). RNA production in the cytoplasm demonstrated using radioactivity; after merotomy the protozoon is maintained in a sterile medium containing the RNA-component ^{3}H-cytidine
   a *Tetrahymena* fragment without nucleus: no production of RNA
   b fragment with nucleus: copious RNA produced

stomata) (figure 78 d). If, after merotomy, a fragment containing a nucleus and a fragment lacking a nucleus are maintained in a sterile nutritive solution with a component of RNA in a radioactive form, ^{3}H-cytidine, RNA formation, readily discerned by its radioactivity, occurs only in the cytoplasm of the fragment containing a nucleus (figure 107 b). As studies with the giant chromosomes of Diptera have shown, the RNA occurs at specific sites on the chromosomes, the nuclear organizers. The nucleoli of the nucleus act as collecting centres for the temporary accumulation of RNA. The basic substance of the nucleoli is protein, and the RNA content usually amounts to only 5–10%. These relationships indicate how important the chromosomal fraction and hence also polyploidy is for the somatic functions of the nucleus.

We have up to now shown the importance of the nucleus with its genes to the somatic functions of the whole cell body, but the reverse is also true: an isolated nucleus cannot survive without its cytoplasm. Nucleus and cytoplasm together form a physiological entity. The reciprocal relationship between nucleus and cytoplasm is demonstrated by the detailed findings of electron-microscopic studies, which show that the nucleus does not just lie in the cytoplasm as a discrete body enclosed in the nuclear wall. Figure 108 shows diagrammatically (bottom left) a partial section through a nucleus, with a nuclear wall of two membranes traversed by pores. Each of these unit membranes has a thickness of about 7–8 nm. The space between the two membranes, the perinuclear space, can be of varying width with an average width of 10–15 nm. The two nuclear membranes and the perinuclear space differ from one another in their response to osmium. Whilst the membranes are osmiophilic, the perinuclear space is osmiophobic.

There are three possible routes for the exchange of substances between the interior of the nucleus and the cytoplasm: an exchange via the pore system, by the formation of vesicles, and via the canal system of the endoplasmic reticulum. The **nuclear pores** with a diameter of 30–100 nm in no way make the nucleus an absolutely open organelle. Studies have shown that an experimental opening of the nuclear wall leads to the death of the cell. The appearance under the electron microscope suggests that the cell pores are equipped with a diaphragm, although the question then arises as to whether such diaphragms are formed only temporarily when the need arises. It has nevertheless been shown that substances of high molecular weight such as serum proteins do not diffuse into the interior of the nucleus; this indicates that a selection process is operating. Just as with any other biological membrane, the nuclear pores are only semi-permeable; even substances with a high molecular weight can pass in the other direction, from the interior of the nucleus into the cytoplasm. In some amoebae (*Amoeba proteus*, *Pelomyxa* sp.), cylindrical ring-shaped thickenings with a honeycomb-like or hexagonal prismatic structure (annuli) are deposited at the pores near the karyoplasm. These structures are also present in *Gregarina rigida*. Similar structures are deposited on the outer nuclear membrane of *Endamoeba blattae* which live in the cockroach.

A further possible way for the nucleus to emit substances into the cytoplasm is through **production of vesicles** by the nuclear wall. This has been demonstrated in the amoeba *Pelomyxa carolinensis*. The nuclear wall protrudes into the cytoplasm, and this bulge is then cut off to form a vesicle, or utricle, which later dissolves in the cytoplasm, thereby releasing its contents.

It is particularly worthy of note that the Protozoa also have

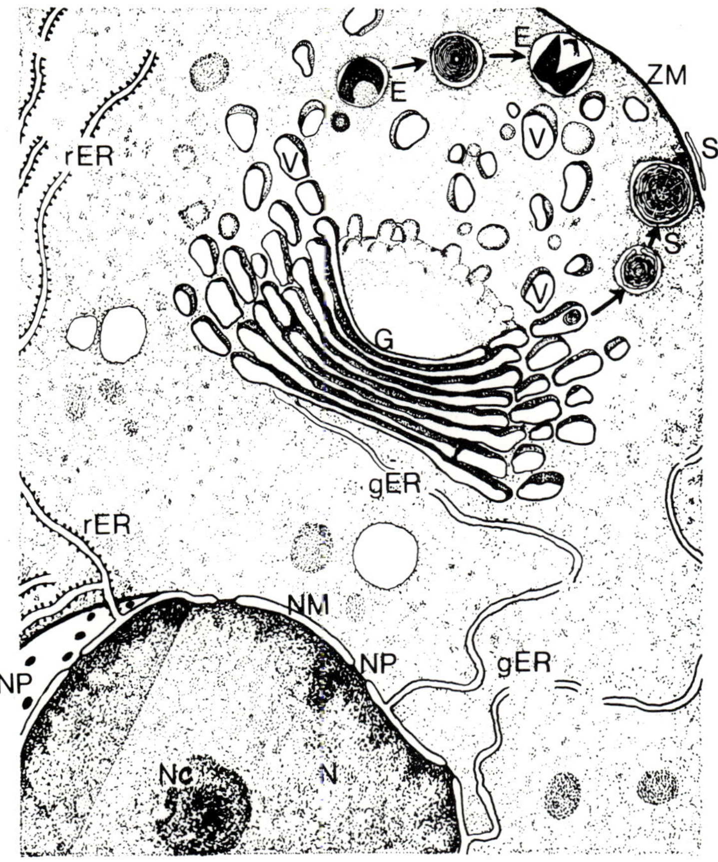

Figure 108. **Nucleus, endoplasmic reticulum and Golgi apparatus.**
(Diagram of the appearance in the electron microscope)

G = basic elements of the Golgi apparatus
V = vesicle
E = ejectisome
S = scales
ZM = cell membrane
N = cell nucleus
NM = nuclear wall with inner and outer nuclear membrane
NP = nuclear pores of the nuclear wall
gER = smooth endoplasmic reticulum
rER = rough endoplasmic reticulum with ribosomes on the outside
Nc = nucleolus

a direct connection between the endoplasmic reticulum and the perinuclear space. The **endoplasmic reticulum** is a fine branched system of canals which extends through the whole cytoplasm. Under the electron microscope it appears tubular in form, or with fine vesicles and flattened sacs, but this variable appearance may be caused by the tubes being cut through in different planes. The membrane of the endoplasmic reticulum contains proteins, phospholipids and abundant RNA. Like the internal nuclear membrane, it operates in the selection of substances which are transported via the lumen of the endoplasmic reticulum from the karyoplasm to the cytoplasm or vice versa.

The canal system of the endoplasmic reticulum, surrounded by a unit membrane, and embedded in the **ground cytoplasm** or **hyaloplasm**, is directly connected with the perinuclear space (figure 108). Two outwardly different forms of the reticulum can be distinguished: in figure 108, smooth endoplasmic reticulum is on the right of the nucleus and rough on the left. There is no essential difference between these two forms. The appearance of the rough reticulum, which has also in the past been called the **ergastoplasm**, derives from the smooth form by the deposition of granules with a diameter of 10–15 nm. These granules, lying close to the reticulum in the ground cytoplasm, are rich in RNA and are called **ribosomes**. They can also lie in the hyaloplasm between the canals of the endoplasmic reticulum, particularly in cells with intensive protein synthesis, since protein synthesis occurs at the ribosomes (Scheme 1, page 187).

Figure 109 shows once again, this time in electron micrographs of *Trypanosoma gambiense* (figure 18 c), the open connection of the endoplasmic reticulum with the perinuclear space. The outer nuclear membrane passes over into the endoplasmic reticulum without interruption. However, this anchoring of the cell nucleus to the endoplasmic reticulum in no way results in an inflexible siting of the nucleus in the protozoan cell. These internal structures are not rigid elements. During locomotion of some species of amoeba, the nucleus can be seen moving around in all parts of the cytoplasm. The siting of the nucleus in the cell body is fixed only in the Protozoa which have internal strengthening provided by fibrillar supporting elements, as in many ciliates. This occurs especially in

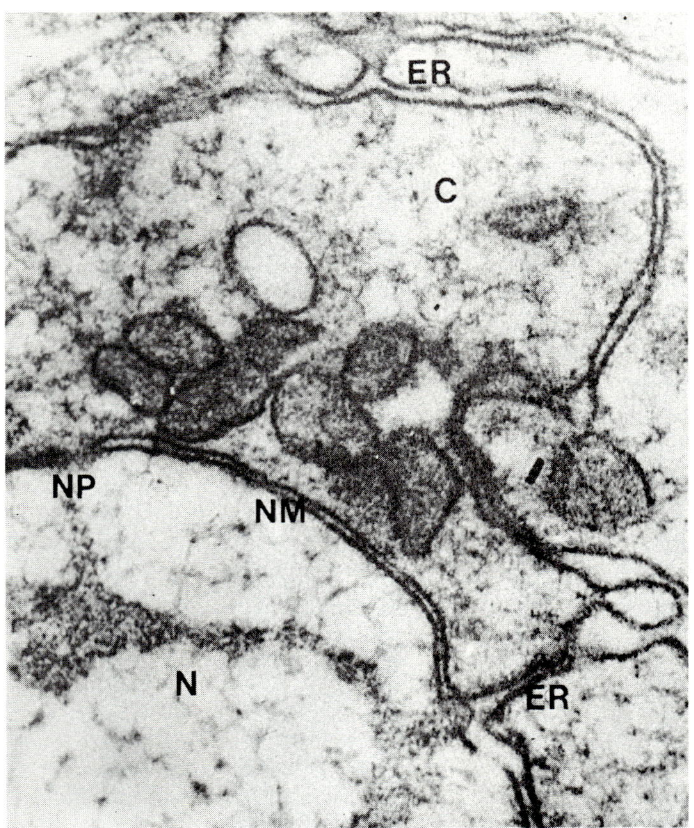

Figure 109. **Nucleus and endoplasmic reticulum** of *Trypanosoma gambiense*

    N = cell nucleus
    NM = nuclear wall
    NP = pore of the nuclear wall
    ER = endoplasmic reticulum
    C = ground cytoplasm (Magnification 65,175 ×)

the Entodiniomorpha (figure 83), which are very much strengthened.

The possession of a nuclear wall, which determines the classification of the Protozoa as eukaryonts, is not universal at all times. The nuclear wall, including the outer membrane which belongs to the endoplasmic reticulum, dissolves during mitosis in most eukaryotic cells, and does not form again until after telophase. This process is directly dependent on the condition of the nucleus and is thus, like all events in the cell, induced by the nucleus. Nevertheless, in some Protozoa, the dissolution process can follow a different course. For example in the amoeba *Amoeba proteus*, only the internal prisms (annuli) dissolve.

Golgi apparatus and parabasal apparatus

In Protozoa and in other eukaryotic cells, the **Golgi apparatus** can be distinguished from the endoplasmic reticulum as a membrane system with its own ultrastructure. Figure 108 shows the characteristic appearance of the cell organelle; its basic elements are small lamellar sacs filled with liquid, compressed at the centre, wider at the ends, bent into bowl shapes and stacked in a pile. This collection of sacs, which are called **cisterns**, usually comprises about 5–8 individual cisterns but in Euglenidae comprises 20–30; the whole constitutes the 0·5–2·0 µm dictyosome. From both ends of this, vesicles of various sizes become detached as transport organelles. Also, large vacuoles can fuse into a group, as is shown in figure 108 on the concave side of the dictyosome. The quantity and size of the Golgi complex differ according to individual protozoan species. The Golgi apparatus is either randomly arranged in the cytoplasm or it may have a specific position in relation to the nucleus. This applies particularly to the Polymastigina, in which osmium fixation reveals the so-called **parabasal apparatus** even under light microscopy.

The normal Golgi apparatus can be detected under light microscopy only after treatment with silver nitrate or osmium tetroxide, because of the resultant precipitation of metal. In contrast, the parabasal bodies are seen more easily and, under electron microscopy, show both the actual membranous Golgi complex and a well-developed parabasal filament, a transversely striated thread of protein running the whole length of the parabasal body (figure 110). The outward appearance of the parabasal apparatus varies; in *Tritrichomonas caviae* it is long and stretched out (figure 19b); in *Devescovina striata* and *Macrotrichomonas pulchra* the parabasal body, which lies anteriorly around the axostyle, ends in a parabasal spiral (figure 20a, b).

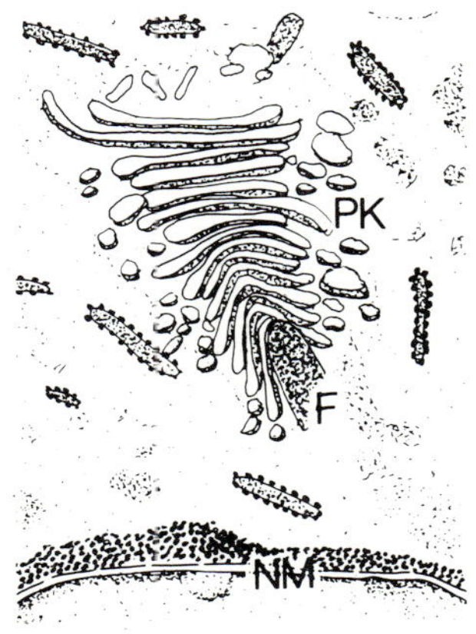

Figure 110. Transverse section of a **parabasal body** of *Trichonympha* (Poly-mastigina, see figure 21)
PK = parabasal body
F = filament of the parabasal body
NM = nuclear wall (Magnification 27,500 ×)

The parabasal apparatus of *Lophomonas blattarum* lies round the nucleus like a crown of rays (figure 20f), while the nucleus of *Trichonympha collaris* is surrounded by large thread-like structures of the parabasal apparatus (figure 21).

The normal Golgi apparatus has been described so far for very many flagellates, for Sporozoa and for various amoebae. In ciliates, the Golgi apparatus is apparently not a constant cell organelle. Thus it has been demonstrated in *Tetrahymena pyriformis* only during conjugation and not in the growth phase. The membranes of the Golgi apparatus correspond in magnitude to the membranes of the endoplasmic reticulum and like the smooth endoplasmic reticulum, they are free from ribosomes. Biochemical studies on centrifuged isolated Golgi substance show that practically no RNA is present. Phospholipids have been demonstrated in the membrane, and there is abundant acid phosphatase, an enzyme of the lysosome granules, which are significant in the assimilation of foodstuffs that have been taken up. A continuity between the dictyosomes of the Golgi apparatus, the endoplasmic reticulum and the lysosomes can clearly be seen in some Protozoa, such as the flagellate *Chrysochromulina chiton*.

The form and function of the dictyosomes of the Golgi apparatus are subject to the control of the cell nucleus. If the nucleus of the amoeba *Amoeba proteus* is removed, the size and number of cisterns in a dictyosome decrease. The significance of the Golgi apparatus for the cell is not yet completely clear. Studies so far indicate that the Golgi apparatus assists in the secretion of proteins and polysaccharides, and in glycoprotein synthesis. In many Protozoa the Golgi apparatus lies in the region of the contractile vacuole, so it may possibly be involved also in the regulation of the water balance in the cell, particularly when the concentration of secretions in the cell is accompanied by an output of water. The light-refractive structures which are deposited as ejectisomes (see also figure 137 e) in the gullet wall of the Cryptomonadina arise in this way, e.g. in *Cryptomonas ovata* and *Chilomonas paramecium* (figure 6 a, b). The scales deposited on the cell membrane of some flagellates, such as the Chrysomonadina, are also produced in the Golgi apparatus and transported by means of vesicles to the surface of the cell (figure 108).

Mitochondria and kinetoplast

Mitochondria also belong to the membranous cell organelles in the cytoplasm. These are the thread granules—granules with a tendency to form threads by association together. They are cell organelles with great plasticity (they measure 0.5–$1.0\,\mu m$ in diameter and up to $10\,\mu m$ in length) which are free to move about in the ground cytoplasm and congregate particularly in regions of the cell which have great metabolic activity. They also tend to accumulate alongside the nucleus, the endoplasmic reticulum or the cell membrane. Vital staining with Janus green B is specific, and treatment with tetrazolium salts gives a very clear picture through the production of formazan, which gives a red-blue colour on account of the enzyme content of the mitochondria.

After the mitochondria have been concentrated in the ultracentrifuge, it is possible to identify them as the site of biological oxidations; they are of central significance in the whole oxidative metabolism of the cell, and especially in cell respiration. The uniform membranous structural basis of all mito-

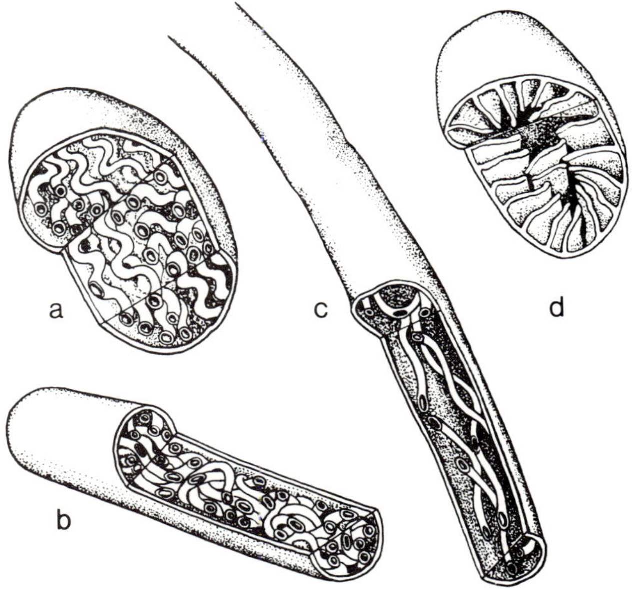

Figure 111. **Mitochondria** from 4 protozoan species (diagrammatic):
  a from *Pelomyxa carolinensis* (amoeba)
  b *Paramecium caudatum* (ciliate)
  c *Trypanosoma lewisi* (flagellate)
  d *Euglena spirogyra* (flagellate)
a–c: tubuli type,
d: cristae type.

chondria was first shown by electron microscopy. Like the cell
nucleus, the mitochondria are surrounded by two unit membranes which are separated by a space of about 7 nm (figure
111). The smooth outer membrane constitutes the boundary
with the cytoplasm, while the inner membrane produces a
marked increase in surface area by means of numerous protrusions into the interior of the mitochondria. In the Protozoa,
this increase in area leads in most cases to the formation of
sinuous tubular processes which run through the internal mitochondrial cytoplasm, the matrix (figure 111 a–c). This form of
mitochondrion is known as **tubuli type**. In the Euglenoidina
and some other flagellates the inner membranes of the mitochondria produce septa which project into the matrix and,

depending on whether they are flat or sac-like, are called **cristae** or **sacculi**. Figure 111 d shows a cristae type.

The mitochondria also contain DNA, which differs, however, from the nuclear DNA in having a more homogeneous base composition and in lacking histones. In this respect this mitochondrial DNA resembles bacterial DNA. The greater homogeneity of the mitochondrial DNA in comparison with the nuclear DNA derives from the smaller proportion of guanine and cytosine which, for example, in *Tetrahymena pyriformis* amounts to only 25%. In the ciliate *Tetrahymena* it can also be shown that multiplication of mitochondrial DNA does not proceed in synchrony with that of the nuclear DNA. The multiplication of the mitochondria themselves is effected by division, as light microscopy shows. However, quantitative considerations lead to the opinion that the DNA in the mitochondria cannot provide sufficient genetic information for the production of new daughter mitochondria. Dependence on the nucleus must thus be assumed for the mitochondria too, in respect of their division.

It has already been said that the mitochondria are at the centre of cell respiration and oxidative processes. The number and structure of the mitochondria give the first indication of the way in which the energy transformation reactions proceed in the Protozoa. *Tetrahymena* has 600–800 mitochondria. This number increases until it is doubled just before the next division. The erythrocytic trophozoite of mammalian plasmodia has only one body with a double membrane which can be regarded as the equivalent of a mitochondrion. The reactions of energy metabolism are determined by the type and amount of the substrate available to the Protozoa as fuel, as well as by the partial pressure of oxygen and carbon dioxide. It is therefore understandable in Protozoa which always live anaerobically (e.g. the intestinal amoeba *Entamoeba histolytica*; the amoeba *Pelomyxa palustris* which lives in stagnant marsh water; the flagellates of termites; and the ciliates in the rumen of ruminants) mitochondria are absent or, at least, the typical internal structure of the mitochondria cannot be detected. In Protozoa which can alternate between the aerobic and the anaerobic way of life, the transition to anaerobic life, with the loss of the oxidative citric acid cycle, is accompanied by the disappearance of the mitochondrial tubuli.

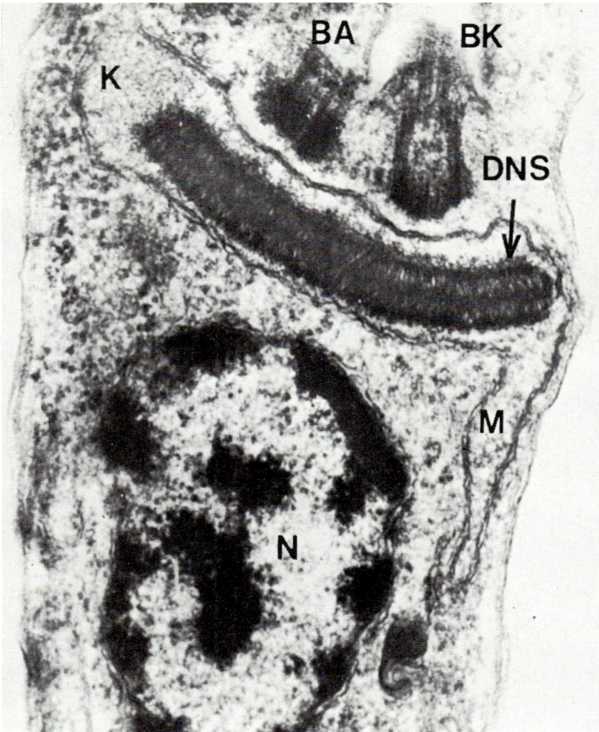

Figure 112. **Kinetoplast** of the culture form of *Trypanosoma cruzi*
  K = kinetoplast
  DNS = DNA of the kinetoplast
  M = mitochondrion
  N = cell nucleus, surrounded by the nuclear wall
  BK = basal body of the flagellum
  BA = site of a new basal body (Magnification 38,910 ×)

The **kinetoplast** of the Trypanosomatidae and Bodonidae
(figure 17 e, f, g), which both belong to the Protomonadina, can
be regarded as a special form of mitochondrion. The kinetoplast
of the Trypanosomatidae is not constant in size and shape, but has
an average size of about 1 µm and always lies in the region of
the basal body at the base of the flagellum. In a Giemsa
preparation the kinetoplast is stained in the same way as the
nucleus (figures 17 b, c, d, 18 c, d). With the Feulgen reaction the
kinetoplast proves to be Feulgen-positive and therefore must

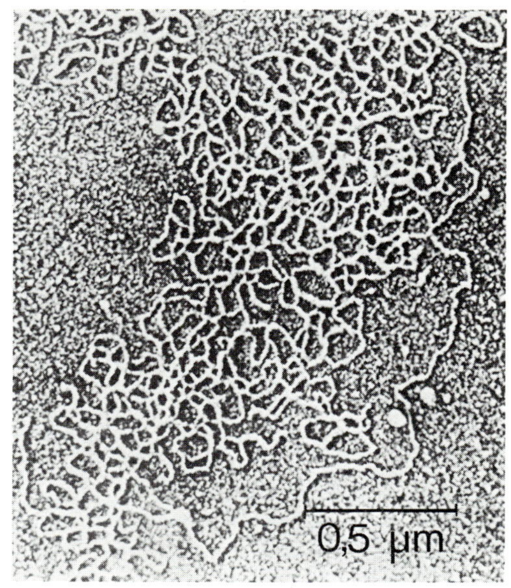

Figure 113. Electron-microscopic view of circularly and linearly arranged molecules of the **kinetoplast-DNA** in *Trypanosoma cruzi* (Magnification 40,000 ×)

contain a certain amount of DNA. Electron microscopy shows that the DNA-containing kinetoplast is only a special constituent of a sac-shaped mitochondrion (figure 112). Only the section of the mitochondrion at the top of the figure has the appearance of a kinetoplast and contains the DNA which is easily detectable. The DNA of the kinetoplast, like that of the normal mito-chondria, contains no histone, in contrast to the nuclear DNA. It can be distinguished from that of the nucleus as the so-called satellite-DNA; it does not produce chromosomal structures, and in the purified state consists of small circular molecules with an individual size of around 0·45 µm (figure 113). In this, the DNA of the kinetoplast is the same as the general mitochondrial DNA. The action of various substances on living Trypanosomatidae renders the kinetoplast Feulgen-negative, and the fibrillar DNA is no longer detectable. Such individuals with atypical kineto-plasts are called **dyskinetoplastic**. In trypanosomes of the African *Trypanosoma brucei* group, which also includes *T. gambi-ense,* the causal agent of sleeping sickness, the development in the blood of the vertebrate host is not impaired by the absence of a normal kinetoplast, although such dyskinetoplastic trypanosomes are no longer viable in the insect vector; this is connected in particular with the effect on the synthesis of respiratory enzymes.

The multiplication of the kinetoplasts is achieved by autonomous binary fission, in which division of the kinetoplast begins before division of the nucleus. In the developmental cycle of the Trypanosomatidae, a stage is usually included in which the kinetoplast lies in close proximity to the cell nucleus. Even structural connections between these two organelles have been seen on more than one occasion. The functional significance of this relative position, and whether exchange reactions are occurring here, is not yet clear.

The plastids

The plastids (chromatophores) can be differentiated according to their colour under the light microscope as green chloroplasts, chromoplasts of other colours, and colourless leucoplasts. The appearance of the plastids in the electron microscope shows a basic correspondence with that of the mitochondria. A matrix, the stroma, is separated from the cytoplasm by a double membrane. The inner delimiting membrane undergoes an increase in active surface area by the formation of lamella-like processes extending into the stroma, the stromatic lamellae (figure 114); each lamella comprises densely packed compressed membrane sacs, the **thylakoids**. Instead of producing continuous lamellae, the thylakoids can in certain places produce textured striped cylindrical grains, the **grana**, which can be seen even under the light microscope and which, however, occur only rarely in Protozoa. On the other hand, zones with a densely granulate or fibrillar structure, the **pyrenoids**, are more frequently encountered in the plastids of the flagellates; starch or paramylum grains tend to be deposited on these rather more than they are elsewhere (figures 114 b, 13 f).

Our ideas are still very hypothetical concerning the molecular structure and molecular function of the thylakoids, which are composed of structural proteins, lipids and chlorophyll, and which carry enzyme systems. However, it appears as though the

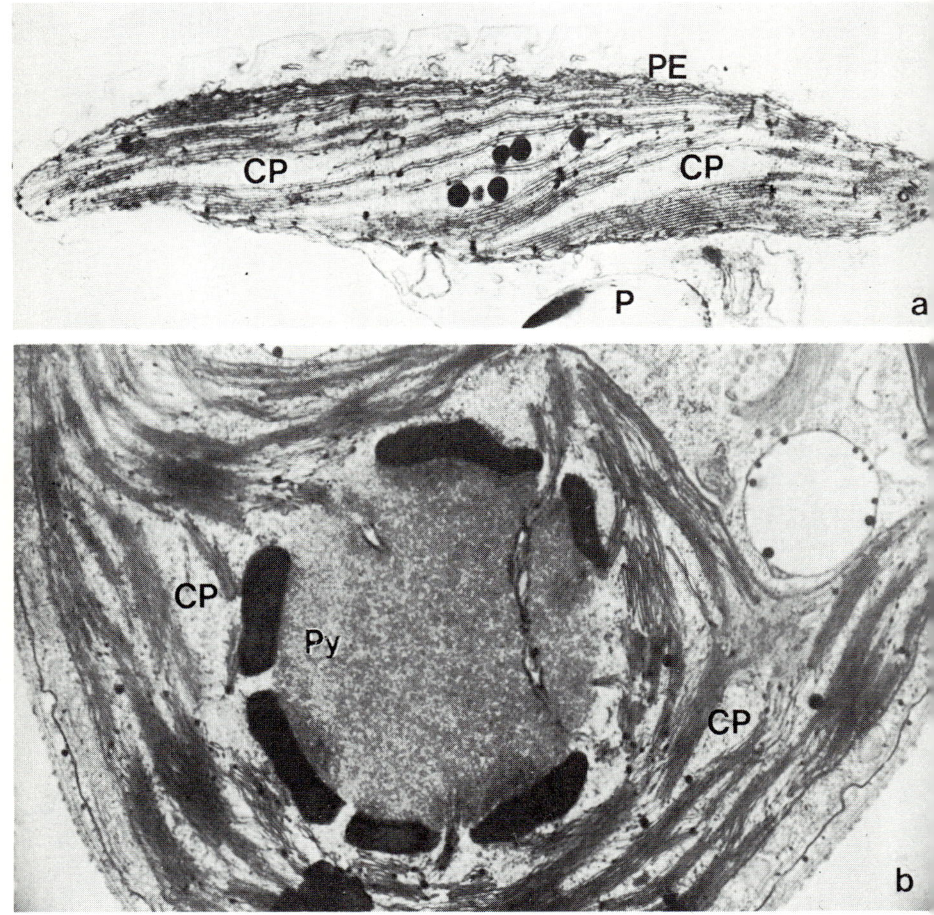

Figure 114. **Chloroplasts**
   a from *Colacium cyclopicolum* (Euglenoidina)
   b *Chlamydomonas reinhardi* (Phytomonadina)
   CP = chloroplast with the stromatic lamellae formed from thylakoids
   Py = pyrenoid, the black deposits on it = synthesized carbohydrates
   P = paramylum granule lying outside the chloroplast
   PE = pellicle of *C. cyclopicolum*
Magnification a = 24,000, b = 19,500.

primary reactions occur in the membranes, and the secondary reactions on their delimiting surfaces and in the stroma. When the active plastids change into the proplastids which form in the dark (figure 22 f, g), the chlorophyll content disappears and the stromatic lamellae composed of thylakoids regress. Under the influence of light, the lamellae are again produced and photosynthesis recommences. It can be deduced from this that the lamellae represent the actual structural site for photosynthesis. Exposure of *Euglena gracilis* to culture temperatures of 32–35 °C, to ultra-violet irradiation, or to treatment with streptomycin and other antibiotics, has led to the production of strains which are not only temporarily colourless but which remain permanently without chloroplasts. They have lost the capacity to restore the chloroplasts. Such strains can be cultured in media containing suitable nutrients for an unlimited length of time.

The plastids and proplastids are capable of division. They are autonomous autoreproductive cell organelles. They are distributed between the daughter cells at cell division and, like the mitochondria and kinetoplasts, the plastids contain DNA. It has been demonstrated in *Euglena* that the DNA in the chloroplasts differs from the nuclear DNA in its base composition. Hence the two types of DNA can be separated from one another in a density gradient. About 4% of the total DNA fraction of the cell is accounted for here by the DNA of the chloroplasts. As to the function of this DNA, this is so far mainly conjecture, although it is known to play a part in the extrakaryotic inheritance in the flagellate *Chlamydomonas rheinhardi* (see page 279).

The pellicle

The **pellicle** forms the outer boundary of the protozoan body, and vital life processes are linked with it. The pellicle participates in ion exchange and generally in selective exchange of substances between the cell and its environment; it is also involved in the detection and selection of nourishment and its uptake by molecular transport, pinocytosis and phagocytosis, and also in the elimination of food wastes. It is concerned in sexual pro-

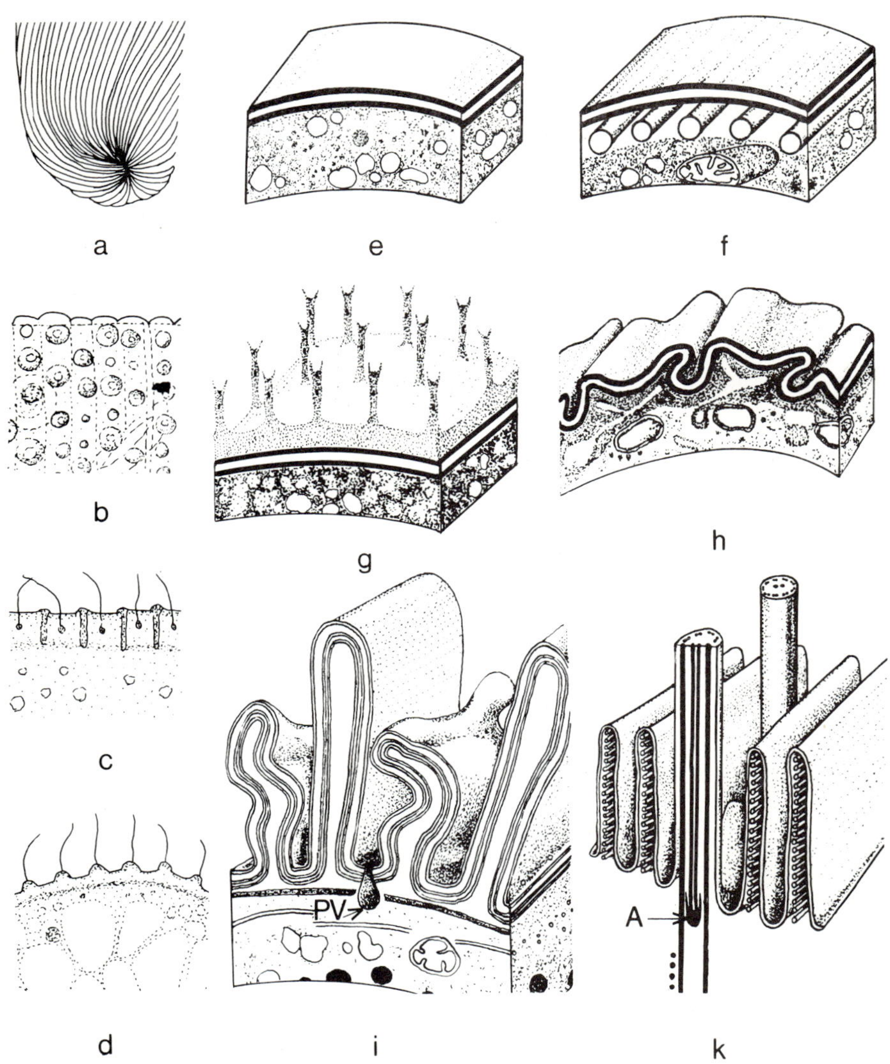

Figure 115. **Pellicular structures**
a–d: light microscopic
e–k: electron microscopic
   a *Euglena ehrenbergi* (flagellate)
   b *Vorticella monilata* (peritrichous ciliate)
   c *Paramecium caudatum* (holotrichous ciliate), transverse section
   d *Loxodes rostrum* (holotrichous ciliate), transverse section

cesses, particularly in the selection of a partner and the ensuing mutual orientation; in cell division; in locomotion; and particularly, in a protective capacity, in cases of mechanical damage and attack by antibodies. A very wide range of pellicle forms is seen, as determined by subpellicular structures. Even under light microscopy, clear differences between the pellicles of different protozoan species can often be detected. While the simple protozoan body, e.g. that of most amoebae, is evidently surrounded by only a thin cytoplasmic skin, other definite formations can be seen. In the flagellate *Euglena ehrenbergi* (figure 115 a) the cell surface appears to be spirally striped. The pellicle of the ciliate *Vorticella monilata* (figure 115 b) has wart-like thickenings. The surface of *Paramecium caudatum* (115 c) has a pattern of angular zones surrounded by prominent ridges, and the cilia originate in the middle of the sunken areas. In contrast, light microscopy clearly shows that the cilia of the protociliate *Opalina ranarum* arise from the rib-like thickenings of the pellicle (figure 115 d). These complex cell coats of the ciliates are also called the **cortex**.

An accurate analysis of the structure of the pellicle is possible only by electron microscopy (figure 115 e–k). Studies of the intestinal amoeba *Entamoeba histolytica* (figure 115 e) show the pellicle to be a closed system of membranes, the **plasmalemma**. Two layers of about 2·5 nm thickness, which show high contrast under the electron microscope, are separated by a lumen of the same width. This arrangement of membranes is so universal that the plasmalemma is also called a **unit membrane**. Primarily lipids and proteins, but also carbohydrates, are involved in its structure. There are as yet no uniform data concerning the molecular structure of the lipoprotein membranes. These consist either of lipoprotein complexes or of two protein layers separated by a lipoid layer. In *Trypanosoma* and *Leishmania* (figure 115 f), a layer of fibrils lies beneath the plasmalemma. In an electron microscope these **fibrils** are seen

---

e *Entamoeba histolytica* (intestinal amoeba)
f *Trypanosoma, Leishmania* (flagellates)
g *Hyalodiscus simplex* (amoeba)
h *Euglena spirogyra* (flagellate)
i *Lecudina pellucida* (gregarine) with pinocytosis vacuole PV
k *Opalina ranarum* (protociliate), A = axial grain of the cilium.

as tube-shaped hardened cytoplasmic structures, the micro-tubules, with a diameter of about 24 µm. They assist in the general stabilization of the cell, linked with a varying ability to contract.

In the amoeba *Hyalodiscus simplex* a peculiarly structured layer, called the **glycocalyx**, is deposited on the unit membrane (figure 115 g). In the flagellate *Euglena spirogyra* (115 h) and *Lecudina pellucida* of the Gregarinida (115 i) the pellicle has various convolutions. Mitochondria can be seen here in the cytoplasm beneath. The protociliate *Opalina ranarum* (115 k), too, has a folded pellicle. The cilia can be seen here to arise from the bottom of the folds (compare figure 115 d), and numerous fibrils pass along the folds. The folding of the pellicle markedly increases the surface area of contact between the protozoan cell and the outside world.

Cuticle and other alloplasmic structures

Whilst the pellicle consists of living cytoplasm capable of division and is thus euplasmatic, the cysts of the Protozoa are surrounded with dead material which is incapable of division and alloplasmatic; this is called **cuticle** to distinguish it from the pellicle. Encystment in Protozoa is achieved by a rounding-off of the cell, expulsion of food wastes and a loss of water, while a strong coating of tectin (pseudochitin) is secreted layer by layer; this is formed from proteins and carbohydrates, just as in the glycoproteins of the true mucilaginous substances, the mucins. Inorganic substances can also occasionally be embedded in it, e.g. silica scales in the cyst wall of *Euglypha alveolata* in the Testacea (figure 116 h). Cyst formation may be induced by nutritional deficiencies, chemical changes in the environment, desiccation or chilling. In gut parasites and the Sporozoa, the cysts assist in transmission of infection. The production of cysts in the form of oocysts and of encysted spores is part of the regular developmental cycle in the Sporozoa (e.g. figure 56).

Most cysts have a cyst membrane which appears as a mainly uniform, thick or thin, cyst wall when seen under the light microscope (see figures 1 b, d, f, 23 l–n, 116 b–g, i, k). Some cysts are clearly multilayered, even under the light microscope, such as the perennating cyst of the ciliate *Colpoda cucullus* (figure

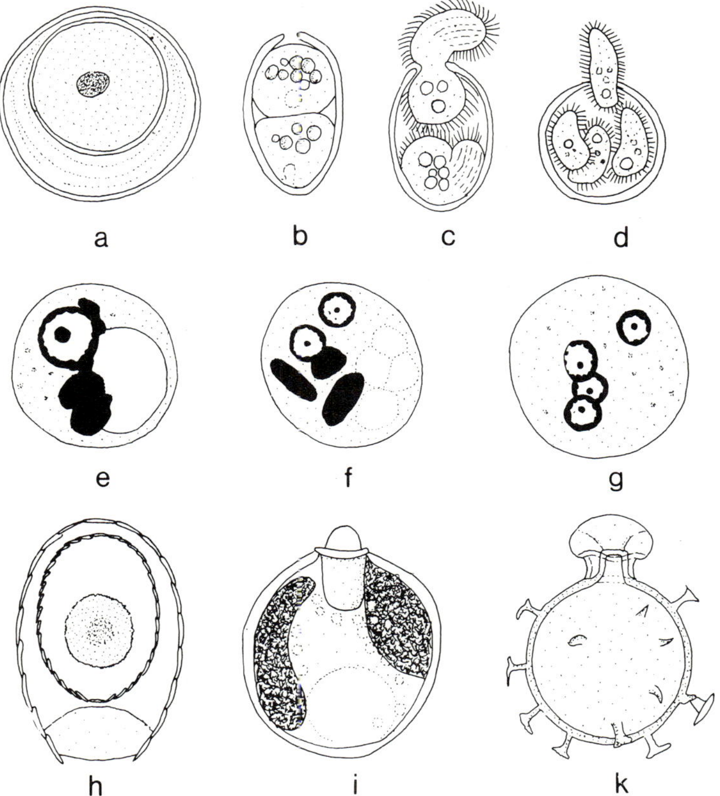

Figure 116. **Cysts**

a multilayered perennating cyst of the ciliate *Colpoda cucullus*

b–d division cysts of *Colpoda cucullus*

  b with binary fission of the ciliate in the cyst

  c emergence after binary fission

  d emergence after division into four

e–g cysts of *Entamoeba histolytica* stained with haematoxylin

  e young uninucleate cyst with glycogen vacuole and chromatoid bodies (reserve substances)

  f binucleate cyst with 3 chromatoid bodies

  g mature 4-nucleate cyst

  h cyst of *Euglypha alveolata*; the ectocyst, covered with silica platelets, lying in the closed Testacean shell; the small endocyst lying centrally

  i cyst of the flagellate *Chromulina freiburgensis*, cyst coat with oral funnel and closure plug; inside are 2 chromatophores, fat droplets and, below, leucosin as a reserve substance

  k prickly cyst of the flagellate *Ochromonas fragilis* with oral funnel and plug

116 a) with an ectocyst and an endocyst. While most cysts have a closed cyst membrane which dissolves under favourable conditions, e.g. for gut parasites in the gut of a new host, others have a preformed opening through which the vegetative stage can slip out. This opening can be closed by a plug in the resting condition (figure 116 i, k). Differences in the formation of the cyst wall are also seen in the electron microscope. In the cyst of *Iodamoeba bütschlii* (figure 29 b) the plasmalemma is covered by a dense 0·05 μm-thick layer, with a 0·2 μm-thick outer layer of lower density. The cyst of *Naegleria gruberi* (figure 17 l) is one of the cysts with a preformed pore, through which the vegetative form can emerge. The cyst wall consists here of an outer irregular coat, roughly 0·025 μm thick and an inner coat 0·2–0·45 μm thick, which are separated by a spongy layer. The pore, which is up to 0·6 μm across, is closed by a plug which has a membranous plate.

Division of the organism occurs in the cysts of some Protozoa. The ciliate *Colpoda cucullus* produces special division cysts for this (figure 116 b–d). Division cysts also occur in the flagellate genus *Chlamydomonas* (figure 23 l–n). The perennating cysts of many intestinal amoebae are at the same time division cysts. In the amoeba *Entamoeba histolytica*, which causes dysentery in man, there is regularly a division into four (figures 1 f, 116 e–g), and in the non-pathogenic *Entamoeba coli* (figure 28 e) even a division into eight.

Many structures enveloping the vegetative stages belong among the alloplasmatic structures; these have already been referred to in the taxonomic section. They include the cellulose covering of most dinoflagellates (e.g. *Ceratum*, figure 5 d) and many colony-forming flagellates (e.g. figure 2 h, l), the deposits in the coating of the Coccolithophoridae containing lime (figure 3 g, h), the capsules of the Tintinnoidea (figure 84), the deposits in the shells of Testacea (figure 36–38), the limestone shells of the Foraminifera (figure 39–42) and the silica skeleton of the Silicoflagellidae (figure 2 o, p) and Radiolaria (figure 34).

The fibrillar systems of the Protozoa

The fibrils, already mentioned in connection with the pellicle (figure 115 f, k), are important in many ways in the cell structure

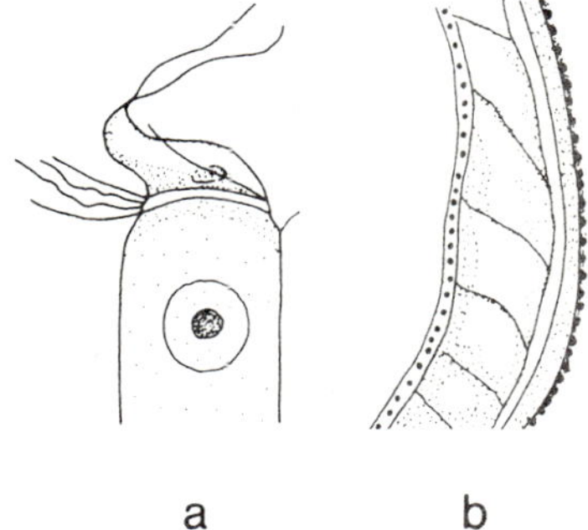

Figure 117. **Systems of fibrils**
a *Giardia intestinalis* with the sucking disc, strengthened by fibrils, applied to an intestinal villus of the small intestine
b *Eudiplodinium medium* (figure 134 b), transverse section of the wall with systems of fibrils in several directions. The rows of dots represent transversely cut fibrils.

of the Protozoa. Intracellular fibrils can easily be seen in some flagellates, even under the light microscope, e.g. as spiral striation or as a border round the cytostome area, as in *Chilomastix* (figure 1 a), or as a border round the concave sucking disc of *Giardia intestinalis* (figure 18 h), which gives sufficient rigidity for sucking on to the gut villi in the small intestine (figure 117 a). The Trichomonadina (figure 19) have a basal fibril which supports the functioning of the undulating membrane. Complex fibrillar systems are formed in the ciliates. In these, several fibrillar systems can be superimposed on each other, particularly if the Protozoa, as in the rumen infusoria which phagocytize large particles of cellulose, are exposed to strong mechanical stress. Figure 117 b shows a section through the periphery of a rumen infusorium. The strong trichites of the basket apparatus in the Gymnostomata are also examples of the fibrillar system of ciliates.

The mechanical strength of the fibrils as they support the cell can be appreciably increased when they are arranged in bundles. Figure 118 a shows an electron-microscopic view of a transverse section through the axopodium of a heliozoon with numerous fibrils associated together as the axis of the pseudopodium **(axoneme)**. Bundles of fibrils are visible in transverse sections of the **axostyles** of the Polymastigina, the Tricho-

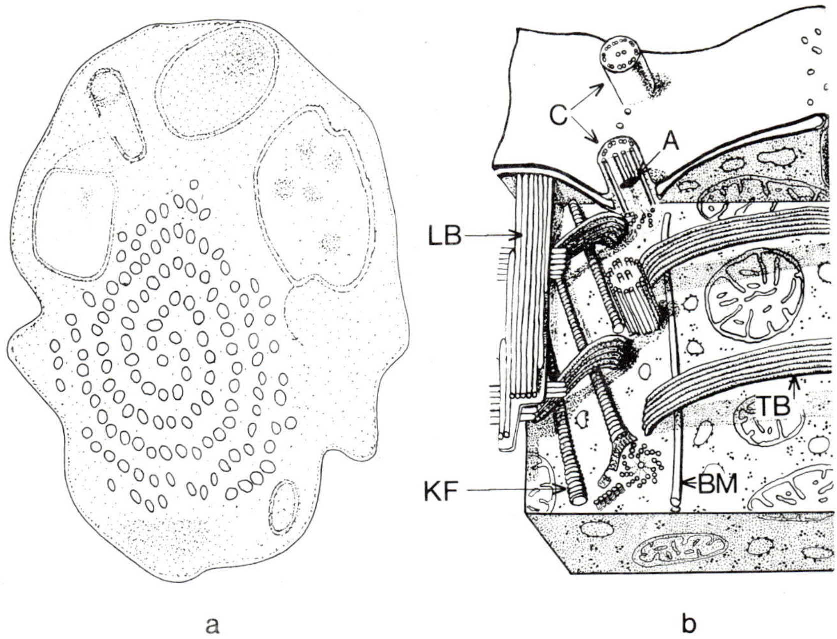

Figure 118. **Systems of fibrils**

a Transverse section through an axopodium of the heliozoan *Echinosphaerium nucleofilum*; at the periphery is the cytoplasm of the pseudopodium; in the centre is the axis (axoneme) formed from numerous fibrils of the microtubuli type

b *Tetrahymena pyriformis*, fibrillar system below the plasmalemma:
    LB = longitudinal microtubular strips
    TB = transverse microtubular strips
    KF = kinetodesmal fibril, BM = basal microtubuli
    C = cilia with the fibrillar system shown in figure 119
    A = axial granule of the cilium

monadina and flagellates of termites (figures 19, 20) when seen in the electron microscope. Figure 118 b shows the electron-microscopic view of *Tetrahymena pyriformis* as an example of the multidimensional arrangement of fibrils in ciliates.

While the fibrils of the axostyle, axoneme and basket apparatus may have an exclusively static significance, other fibrils are characterized by a strong contractibility whose effect can also be enhanced by the bundling together of the fibrils. These fibrils are called **myonemes**. They are present, for example, at the place where the skeletal needles of the Acantharia emerge (figure 35 b), but are particularly striking and widespread in various ciliates. Among the heterotrichous ciliates, the genera *Spirostomum* and *Stentor* (figure 82) have strong contractibility. The peripheral fibrils are here easily visible as striations, even under the light microscope. The Folliculinidae, Tintinnidia and loricate Peritrichidae can draw themselves back into their capsules by means of the myonemes (figures 82, 84, 89). But the aloricate peritrichous Vorticellidae are also largely contractile; here, the stalks suddenly contract spirally or in a zig-zag form, and then slowly extend again. This reaction is controlled by the stalk muscles, i.e. the myonemes, which can be seen in the stalk cytoplasm even under the light microscope, but especially under phase contrast (figure 88 a, e, f).

The locomotory organelles

The locomotory organelles aid systematic classification; they include the flagella and cilia, which also belong to the fibrillar systems. Both appear as fine thread-like oscillating organelles under the light microscope. The flagella of the flagellates generally equal or exceed the cell in length and, except in the highly specialized Polymastigina, they occur only in small numbers. Flagella occur not only in the flagellates but also in the gametes of the Foraminifera and Sporozoa, and the swarmers of the Radiolaria. Cilia are shorter in relation to the cell and are usually numerous, but the latter does not hold for all ciliates. There is no clear line of demarcation between a flagellum and a cilium with regard to relative length. Even the fine structure seen in the electron microscope does not reveal any fundamental difference between flagella and cilia. Both have the same structural arrangement, which can be seen in figure 119 and which applies for all flagella and cilia of other eukaryotes.

The free portion of the flagellum, the shaft (G), is surrounded, with its ground cytoplasm, by the pellicle (P). The effective fibrillar system lies embedded in the ground cytoplasm; there are two central longitudinal fibrils (ZF) and nine pairs of peripheral double-fibrils (PF) in which one fibril of each pair is equipped with two lateral processes, the arms, as can be seen in the transverse section b in figure 119. Each of these 20 longi-

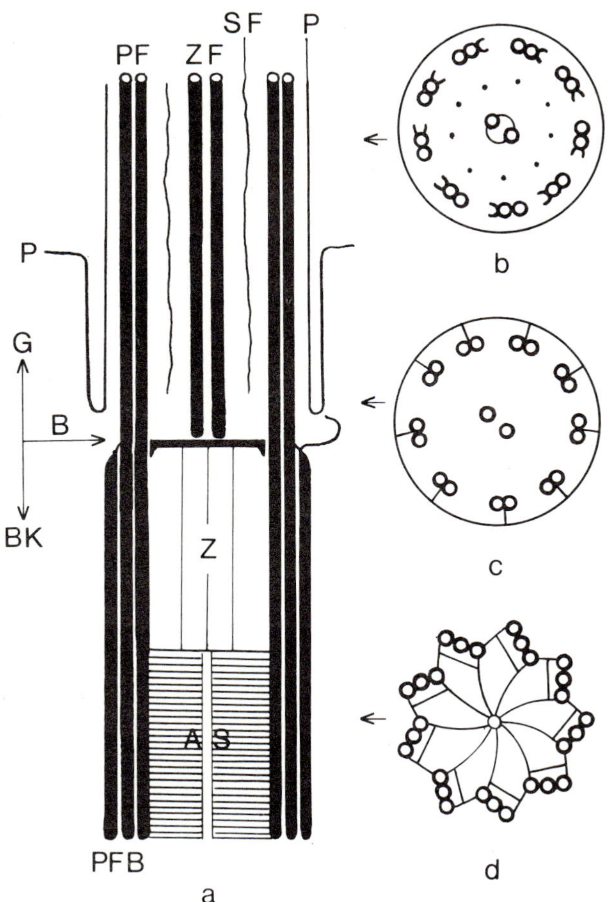

Figure 119. Diagram of the **structure of a flagellum**
a longitudinal section
b–d transverse section
    a G = flagellum above the basal plate
    B = basal plate
    BK = basal body
    ZF = central fibrils
    PF = peripheral fibrils
    SF = secondary fibrils
    P = pellicle (cell membrane)
    Z = cylinder
    PFB = triple peripheral fibrils of the basal body
    AS = spokes of the basal body

tudinal fibrils of the microtubule type in the flagellum has a diameter of about 15 nm. Between the 2 central fibrils and the 9 pairs of peripheral fibrils are an additional 9, markedly thinner, simple secondary fibrils (SF). The two central fibrils end basally at the basal plate (B), which can be developed as an axial grain (axosome) in ciliates (A in figure 115 k and 118 b). Below the basal plate, the flagellar shaft changes into the basal body (BK), in which there are no central fibrils. The basal body of ciliates is also often called a **kinetosome**. The nine peripheral pairs of fibrils are now usually strengthened by an additional fibril. These fibrils are arranged obliquely in the cross-section of the basal body, and the fibrillar arms are no longer present. In the centre of the basal body underneath the basal plate lies the cylinder (Z). Sections from lower down can include spoke-like structures radiating from the middle (AS). There can be variations in the structure of the flagellar shaft, and especially in the basal body in different Protozoa. For example, fibrillar anchorage granules can occur peripherally in the portion of sunken flagellum above the basal plate. The basal body is open to the cytoplasm. It is retained even in the transient amastigote stage of the flagellates, and the new flagellum is formed there. Divisions of the basal body have never yet been observed. It is assumed that the site of a new basal body is determined by those already present, especially since it is always just near an existing basal body (figure 112 BA near BK).

Various surface structures can be seen in the flagella of the flagellates, even under the light microscope, after specific staining or in fresh preparations in a dark field. Whilst the thread-like flagella are uniformly smooth up to the end of the shaft (figure 120 a, b, right flagellum), some flagellate species have flagella with hair-like processes, the **mastigonemes**. Flagella of this type are known as ciliary flagella or **Flimmergeissel** (figure 120 a, b, left flagellum). The arrangement of the ciliary structures (flimmer) is typical for each systematic group. In the Euglenoidina the flagellum has a single line of ciliary

---

b transverse section through the free flagellum with 2 central fibrils, 9 secondary fibrils and 9 pairs of peripheral fibrils, in which one fibril of each pair has two arms. Outside the cell membrane

c Transverse section of the sunken flagellum just above the basal plate

d Transverse section of the basal body in the region of the spokes

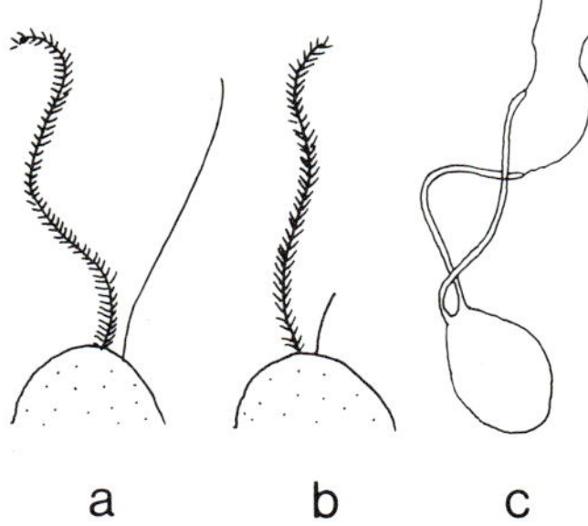

Figure 120. Various **forms of flagella**
  a *Synura* sp. (figure 2g) with ciliary flagellum and thread and thread flagellum
  b *Monas socialis* (Chrysomonadina) with ciliary flagellum as the main flagellum and the thread flagellum as subsidiary flagellum
  c *Polytoma uvella* (figure 12 c) with lash flagella

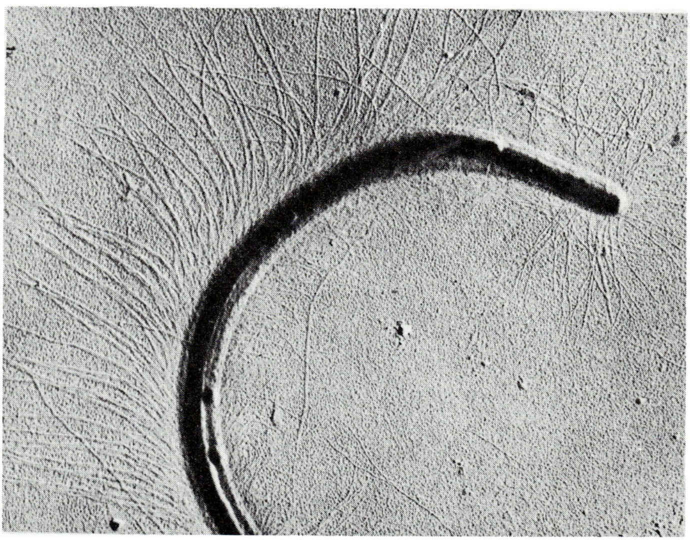

Figure 121. Electron-microscopic view of the ciliary flagellum of *Euglena spirogyra* (figure 10b) (Magnification 9975 ×)

structure wound spirally round it (figure 121), and in the Chrysomonadina they occur all over the flagellum. If a member of the Chrysomonadina has another flagellum, this is a thread flagellum (figure 120a, b). A further variation in flagellar structure is the lash flagellum, in which the peripheral layer of the flagellar shaft is shorter than the central core (figure 120c).

In flagellates with several flagella, the size and arrangement of the flagella can also vary. When they differ in size, the longer is called the **main flagellum** and the smaller the **subsidiary flagellum** (figures 120b, 2k, l, m, 3d, 4b, etc.). Flagella which are directed backwards are called **trailing flagella** in contrast to the **swimming flagella** (figures 8b, p, r, 11a, c, d, f, 17f, g, etc.). For the dinoflagellates, we refer to a **longitudinal flagellum** and a **transverse** or **girdle flagellum** (figure 5d). The trailing flagellum becomes a **ribbon flagellum** when it is equipped with a wide margin of cytoplasm (figure 20a, b). When a flagellum running the length of the cell body is joined to the cell surface it forms an **undulating membrane**. This is formed from a forward-pointing swimming flagellum in *Trypanosoma* and from a trailing flagellum in the Trichomonadina (figure 17d, 18c and 19b–e).

The direction of movement of the flagellate is determined by the varying type of motion of the flagella. Figure 122 shows this using the example of a forward-pointing swimming flagellum. A rapid down-stroke of the mainly outstretched flagellum, followed by a re-erection of the flagellum when bent, results in a movement of the cell forwards (figure 122a). Undulating wave-like beating to and fro results in a movement of the cell backwards (figure 122b), and undulating to one side results in a movement of the cell sideways (figure 122c). The wave-like motion of the flagellum is reinforced by the undulating membranes.

Adenosine triphosphate (ATP) participates in the generation of energy necessary for the motion of the flagella or cilia. The arms of the nine outer peripheral fibrils are probably the site of adenosine triphosphatase activity (figure 119b). An adenosine triphosphatase known as **dynein** has been isolated from the ciliate *Tetrahymena pyriformis*. The outer membrane is apparently not necessary for motion; cilia lacking a membrane can be induced to resume beating by the application of ATP. On the other hand, in the flagellate *Chlamydomonas reinhardi*,

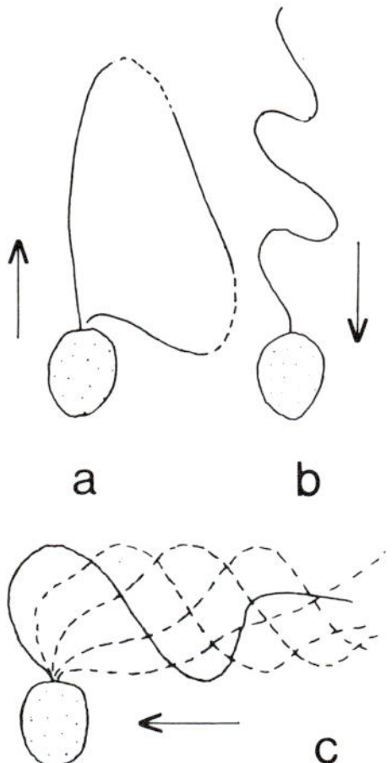

Figure 122. Flagellar beat in *Monas* sp. (Chrysomonadina)
  a when swimming forwards
  b when swimming backwards
  c when proceeding sideways

the removal of the two central longitudinal fibrils by means of proflavin is linked with a loss of flagellar activity. Despite these isolated findings, a thoroughly satisfactory elucidation of the motion of flagella and cilia on a molecular basis is not yet possible.

The type of motion of cilia basically corresponds with that of flagella. In cilia, too, a forward thrust to the cell is achieved by a down-stroke in a mainly extended condition, followed by a re-erection of the flexed cilium. If the ciliature is over-all, as in the holotrichous ciliates, cilia are arranged in parallel rows. Within each row, the continuity is broken only by the suture of the cilia rows (figure 123 a). The cilia which stand one behind another in a row beat metachronously, one after another, whilst each cilium beats synchronously with those at the same level in the neighbouring parallel rows. Thus the extension and flexion of the cilia during the course of locomotion leads to a rhythmical wave pattern like a cornfield in the wind.

The arrangement of rows of cilia in the pellicle is determined by their anchorage in the surface structure. Figure 115 c has already shown, for *Paramecium caudatum*, the position of the cilia

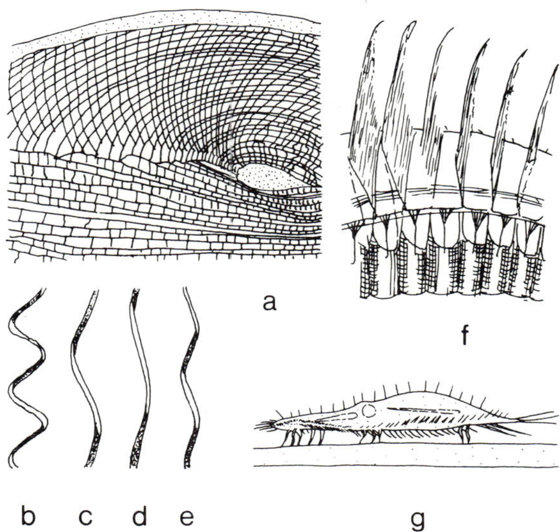

Figure 123. **a** Pattern of the surface of *Paramecium caudatum*, ventral surface with ridge structures and the cilia fields formed by them, and also the suture of the cilia rows with the entrance into the vestibulum.
b–e spiral tracks of the swimming locomotion of ciliates
  b *Strombidium* sp.
  c *Frontonia leucas*
  d *Paramecium aurelia*
  e *Euplotes patella*
  f Membranellae from the zone of membranellae in *Stentor*
  g *Stylonychia mytilus* (figure 86 f) "running" on **cirri**

in the middle of the area between the ridges of the pellicle. A surface view of *P. caudatum* is shown in figure 123 a, where rows of cilia are interrupted only by the ventral suture in which the entrance to the vestibulum is also situated.

The cilia, anchored with their basal bodies in the centre of the cilia fields, are connected with the fibrillar systems of the pellicle. This has already been shown in figure 118 b, using *Tetrahymena pyriformis* as the example. Fibrils come out laterally from the basal body; these are transversely striated when seen in the electron microscope and are called the **kinetodesmal fibrils** (figure 118 KF). All of these kinetodesmal fibrils run forwards in *Paramecium*. They end at the fourth or fifth cilium-field, but overlap the kinetodesmal fibrils of the neighbouring cilia, so that all of them together look like longitudinal fibrils. It is presumed that these fibrils can assist in the co-ordination of the beating of the cilia. In addition to the kinetodesmal fibrils, other fibrils with a simple microtubular structure or transverse striation can come into contact with the ends of the basal body, so that highly complex fibrillar structures in the pellicle can be seen in places, as is evident from figure 118 b.

The arrangement of cilia rows on the surface of the cell, combined with the usually mainly cylindrical, more or less flattened, or even irregularly-shaped cells of the ciliates, effects a swimming locomotion on a spiral track whose convolutions differ according to species. Figure 123 b–e shows some examples of these spiral tracks. When neighbouring cilia stick together in cilia plates, they form undulating membranes, the **membranellae** (figure 123 f). They are particularly typical for the adoral zone of membranellae in the spirotrichs, but membranellae also occur in the holotrichous Hymenostomata and in the peritrichs. These are usually used not so much for locomotion as for generating a current of water which will carry in the foodstuffs. In the formation of membranellae, there is a strong fibrillar and amorphous union of the basal bodies, and a loose connection of the cilia shafts in the region of the basal bodies. Membranellae plates standing parallel to one another beat metachronously like the cilia of one cilia row (figure 123 f). While fusion of the cilia to form membranellae increases the force of the beating motion, association of the cilia into styloid bundles increases their mechanical stability and carrying capacity; these are called **cirri** and function primarily as ambulatory organelles. They are characteristic of the hypotrichous ciliates (figures 123 g, 86). Their movement is by no means so co-ordinated as that of normal cilia or of membranellae; they tend to move more independently of one another.

Cytoplasmic streaming, with the participation of the pellicle, can lead to locomotion. The resultant movements of the cell in free-swimming forms do not, however, invariably bring about a change of locality. In the Euglenoidina, cytoplasmic bulges pass backwards over the cell body like waves (figure 124 a). This process is called **metaboly**. A forwards movement of the whole cell results only when contact is made with a firm substratum. Flowing locomotion is specific particularly to the amoebae.

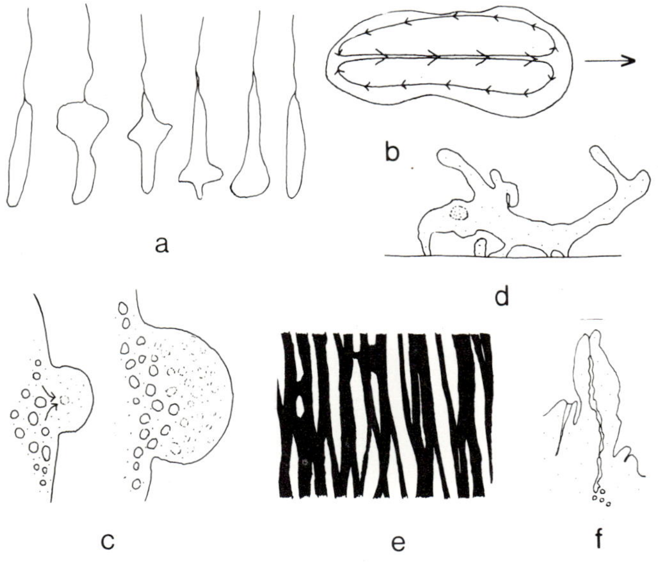

Figure 124. **a Metaboly** in *Peranema* (Euglenoidina).
b cytoplasmic streaming during the forwards-flowing locomotion of *Entamoeba histolytica*
c formation of a hernial-sac pseudopodium in *Endamoeba blattae* from the gut of the domestic cockroach
d stepping locomotion of the amoeba *Amoeba proteus*
e fibrils in the pellicle of *Gregarina polymorpha* (figure 49a) (Magnification 4500 ×) seen in electron microscope
f pinocytosis canal in *Amoeba proteus* with formation of pinocytosis vesicles

Figure 124b shows cytoplasmic streaming during directional movement in *Entamoeba histolytica*. This type of locomotion is best observed with an inverted microscope, i.e. viewed from below. The active cytoplasmic streaming flows in the cell interior as an axial current from the back to the front. The forwards flow of the central cytoplasm causes the peripheral cytoplasm to flow passively to the posterior end. This is linked with a transition from endoplasm to ectoplasm at the front of the amoeba, and from ectoplasm to endoplasm at the back. This process is accompanied by morphological changes in the cytoplasmic structure. Under the light microscope the ectoplasm appears hyaline, glassy-transparent, whilst the endoplasm is granular

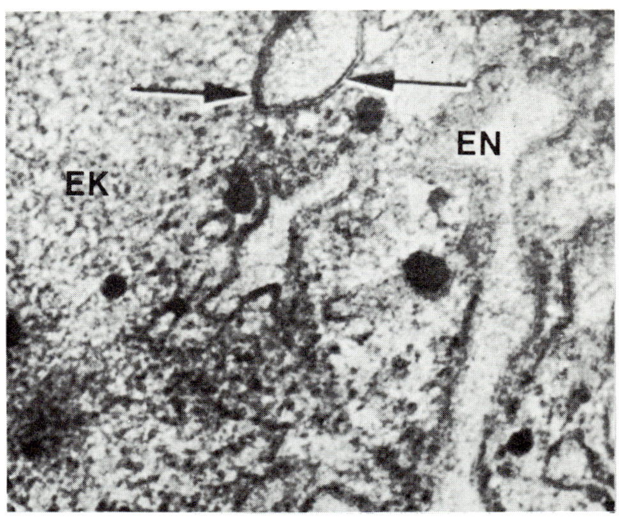

Figure 125. Electron-microscopic structure of the cytoplasm of the amoeba *Hyalodiscus simplex*
Left: ectoplasm
middle and right: endoplasm with pinocytosis vesicles and clear plasma-lemma structure (arrows) (Magnification 73,500 ×)

and vacuolated. Figure 125 shows the electron-microscope differences in structure between the two cytoplasmic zones in the free-living amoeba *Hyalodiscus simplex*; on the left is the denser fine-structure of the ectoplasm, and in the middle and on the right is the more diffuse endoplasm with numerous pinocytotic vesicles which allow the typical cell-wall structure to be seen at the places marked with arrows. Phagocytosis vacuoles and mitochondria can also be embedded here.

In most amoebae the ectoplasm produces branches of a particular formation, the **pseudopodia**. These mobile ecto-plasmic processes can be more or less broad **lobopodia**, rounded at the front (figures 26, 27 a, b, 28 a, b), or slender **filopodia**, pointed at the front (figures 27 c, d, 28 c, d). While the filopodia consist of ectoplasm throughout, the lobopodia soon fill up inside with endoplasm which streams in. However, cytoplasmic streamings are possible not only into a pseudopodium but also in a backwards direction. These cause the pseudopodium to withdraw gradually back into the cell. Lobopodia occur not only in the naked amoebae but also in the Testacea and some flagellates (figure 36, 37 a–d and figure 2 i); the same is true for the filopodia (figures 37 g, h, 38 a, b and figure 2 c, q, 4 c). The production of hernial-sac pseudopodia is a particular kind of lobopodium production. It occurs, for example, in *Entamoeba histolytica*. A local liquefaction causes the ectoplasm to swell out in scattered places to form sacs, until the endoplasm also starts to flows out here (figure 124 c). Close observation of *Amoeba proteus* has shown that the pseudopodia of some amoebae can occasionally be used as ambulatory organelles. Figure 124 d shows this type of stepping locomotion.

Contractions are also evidently involved in the function of the pseudopodia as locomotory organelles; when the pseudopodia have found an attachment in front, the remainder of the cell body is pulled up after them. This is also true for the finely branched reticulate rhizopodia or reticulopodia of the Radiolaria and Foraminifera (figures 34, 42). However, the flowing motion of the amoebic cytoplasm is also evidently caused by contractile elements. Electron-microscopic studies in conjunction with suitable techniques have shown microfibrils running in bundles through the cytoplasm of free-living amoebae such as *Amoeba proteus*, *Chaos carolinensis* and *Acanthamoeba* spp. and also of the parasitic *Entamoeba histolytica*. In *E. histolytica* these complexes of microfilaments are 3–4 µm long and 1–2 µm wide; the individual filaments themselves have a diameter of 2–6 nm. They are absent if there is no cytoplasmic streaming at the moment of fixing. It is assumed that the contraction first affects the forward-flowing cytoplasm, so that this temporarily assumes a solidified gel structure.

In free-floating forms the rhizopodia of the Radiolaria and Foraminifera, like the axopodia of the Heliozoa (figures 32, 33), act as floating processes, corresponding to the floating spines of *Globigerina* (figure 41 f). The capacity of the Radiolaria to float is, however, also determined by their specific gravity. The carbon dioxide synthesized in the metabolism of the Radiolaria lowers the specific gravity of the cell at night by the production of vacuoles, so that the cell rises upwards. In daylight the carbon dioxide is used up by symbiotic zooxanthellae in assimilation processes; the Radiolaria, weighed down by their silica skeletons, sink down again. This vertical movement can amount to 200–350 m a day. Bubbles of gas have also been observed

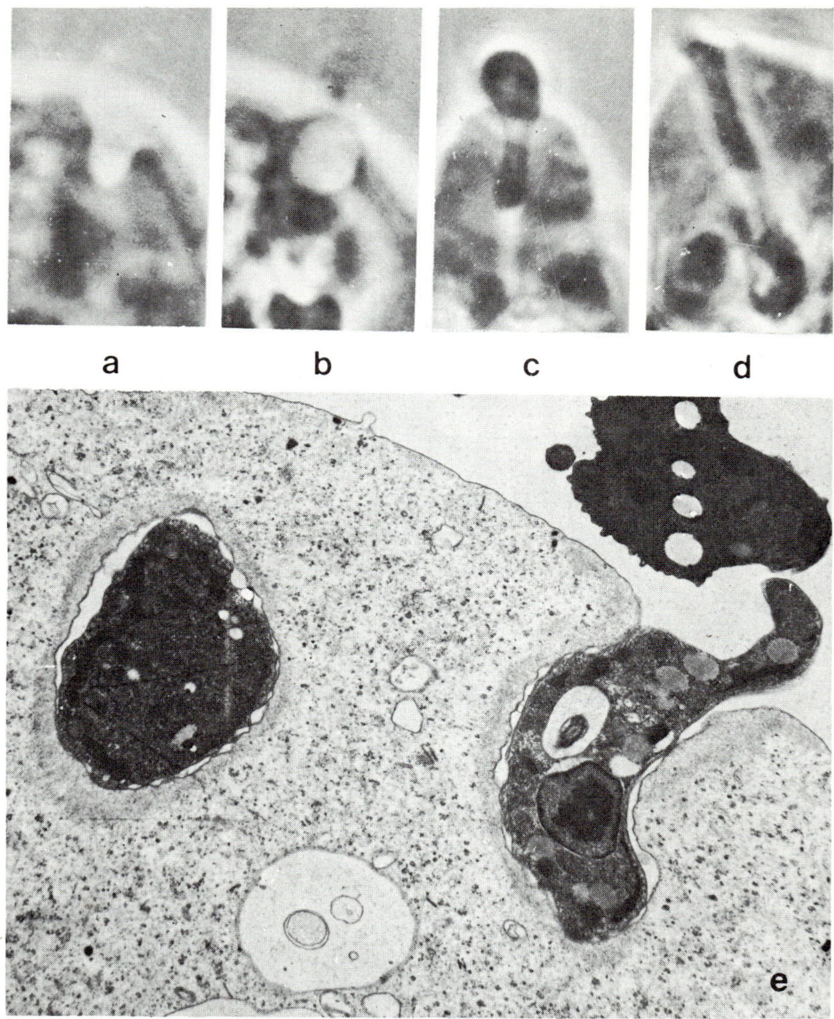

Figure 126. **Pinocytosis** and **phagocytosis** in *Entamoeba histolytica*
a–d phase-contrast view under light microscope. Fixative: glutaraldehyde,
embedded in epon mixture (Magnification 2350 ×)
   a peripheral pinocytotic invagination
   b pinocytotic vacuole already closed
c–d: 2 stages of phagocytic invagination. Uptake of a *Crithidia* cell
   e phagocytic invagination seen in the electron microscope (Magnification
8850 ×)
Above right: section through a *Crithidia* still lying free

acting as organelles for buoyancy in some Testacea of the genera *Arcella* and *Difflugia*.

The Gregarinida have a particular method of locomotion. A gliding motion can be seen under a normal light microscope, although actual organelles for locomotion cannot be discerned. Whereas it was formerly supposed that the Gregarinida secreted a slime which pushed them forwards as it gushed out, it now seems more probable that locomotion is achieved by means of very fine contractions of the pellicle. Numerous fibrils are embedded in the pellicle and these could account for this type of locomotion (figure 124 e). Undulating movements in the cell surface can occasionally be seen, even under the light microscope using time-lapse photography.

Food uptake

The locomotory organelles are also employed in food uptake, by which the protozoan cell procures the raw material for structural processes and energy metabolism. Food uptake can proceed by permeation through the pellicle, by pinocytosis and by phagocytosis. The capacity for **permeation** has already been alluded to in the section dealing with the pellicle (page 209). As a unit membrane, the pellicle carries out a selective exchange of dissolved substances, in accordance with its chemical and physical constitution; in the course of this, smaller molecules and ions are selectively taken up. The simplest form of permeation is diffusion, a passive uptake of substances. In contrast to this, energy is needed for the uptake of matter by permeation in most cases. The molecules which are to be transported are firstly bound to specific sites on the cell surface, then passed through the membrane, and finally freed on the inner side of the membrane. Substances known as **carriers** are involved in this type of transport. These are protein compounds called **permeases**. This function can apparently only be fulfilled with molecules up to a certain size which has not yet been determined. Larger particles of food are taken up in a dissolved

---

below: phagocytosis with formation of a finely reticulate phagocollar
Left: food vacuole with a phagocytozed *Crithidia* in the lumen and with peripheral cytoplasm still forming a phagocollar

state by pinocytosis and as solid food by phagocytosis.

For **pinocytosis**, or "cell drinking", the cell produces invaginations of various forms into the cytoplasm. In *Entamoeba histolytica*, rounded vacuoles arise first (figure 126 a, b) and then migrate into the cell interior. In the free-living amoeba *Amoeba proteus* small vesicles are cut off in the cell interior from the ends of thin funnel-shaped invaginations (figure 124 f). These vesicles, enclosed in a membrane, which arise in pinocytosis can often be seen in large numbers in the endoplasm of amoebae; they can be recognized by the unit membranes which surround them (figure 125, arrowed). Pinocytosis can be stimulated by factors external to the amoebae, particularly by proteins and aminoacids; in the course of this, other dissolved substances which themselves do not induce pinocytosis, e.g. carbohydrates, are also taken up. The extent of pinocytotic activity is, however, also dependent upon the physiological condition of the amoeba, from which it can be inferred that pinocytosis is subject to certain control mechanisms. Although amoebae can produce pinocytotic vesicles anywhere on the cell surface, many Protozoa can obviously only do so at certain places on the surface. When ferritin has been added to the culture medium of the frog trypanosome *Trypanosoma mega*, electron microscopy shows that the ferritin particles only enter the cell at certain places on the flagellar sac near the base of the flagellum. In *Opalina*, pinocytosis occurs at the base between the folds of the pellicle (figure 115 k). This is also true for the gregarine *Lecudina pellucida* (PV in figure 115 i).

The uptake of solid food by **phagocytosis** is easy to see, even under the light microscope, and has therefore been studied in detail for many years. In amoebae which produce vigorous extended lobopodia, two pseudopodia can flow round a food particle and then unite with one another distally, so that a food vacuole is formed (figure 127 a). In other cases, phagocytosis by invagination may occur. An example of this is provided by the uptake of *Crithidia* (flagellates) in a bacteria-free culture of *Entamoeba histolytica* (figure 126 c–e). The pellicle of the amoeba invaginates at the point of contact, as in pinocytosis, and the *Crithidia*, which first of all usually become immobile, are distorted and sucked in (figure 126 c, d). When the invagination canal closes distally, a food vacuole is formed within the cell of the amoeba. Electron-microscopic studies of this process show

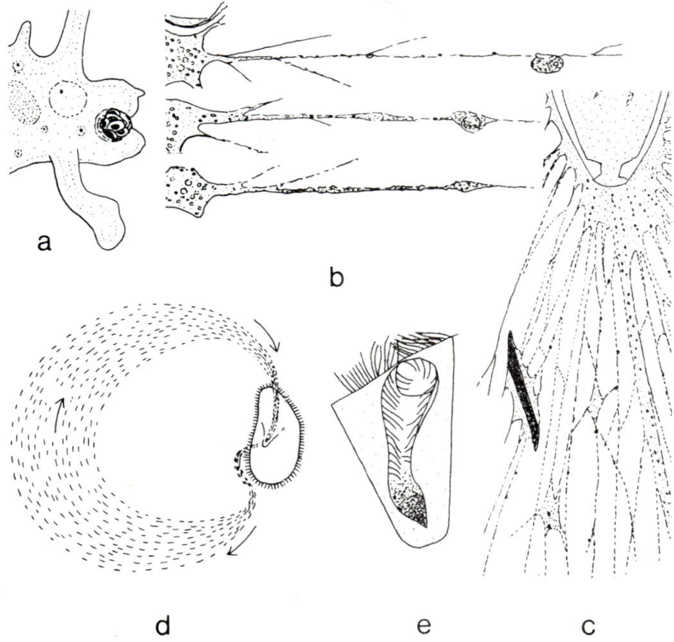

Figure 127. **a** *Amoeba proteus*, formation of a food vacuole by flowing round the food

   b rhizopodium of *Lieberkühnia wagneri* (Testacea), *above* with trapped ciliate, *below* showing phases in the dissolution of the ciliate

   c rhizopodial net of *Gromia ovoidea* (Testacea) with trapped diatom and several rhizopodia which have flowed round it

   d production of a vortex by the peristome ciliature of *Paramecium putrinum*, demonstrated with coloured particles

   e vestibulum of *Carchesium* sp. (peritrichous ciliate) with coloured particles which have been washed in; shortly before the food vacuole is cut off.

that a finely reticulate cytoplasm forms around the invagination canal; this is called the **phagocollar** and is evidently involved in the mechanism of invagination. If a food vacuole is formed (figure 126, left), the phagocollar remains present only for a while. Digestion within the food vacuole begins after the phagocollar has dispersed. The vacuole, like the pinocytotic vesicles, is surrounded by the unit membrane of the pellicle.

The process of phagocytosis, which has been studied particularly in the naked amoebae, also occurs in a corresponding form in the Testacea which have shells, and in many heterotrophic flagellates. Material well suited to the study of this is a culture of *Trichomonas vaginalis* (figure 19 d) after the addition of powdered rice starch, although the swimming form of this trichomonad does not produce pseudopodia. In the Hypermastigidae (figures 20 f, 21), which are related to the Trichomonadidae, and which live in xylophagous termites, relatively large particles of wood are taken up at the posterior end of the cell. Free-living amoebae, too, can phagocytoze and enclose in vacuoles large food particles such as algal filaments which are considerably longer than the body of the amoeba. In the remaining Rhizopoda, some Testacea, the Heliozoa, Radiolaria and Foraminifera, food is trapped by rhizopodia and axopodia. After contact with the pseudopodia, the animal prey are usually rendered immobile by a toxic action, as are the *Crithidia* during uptake by *Entamoeba histolytica*. The trapped prey are subsequently transported to the cell interior by cytoplasmic streaming or contraction of the cytoplasm (figure 127 b). Occasionally the prey begin to dissolve during transport along the pseudopodium. When larger objects are caught, the enveloping cytoplasmic mass can be augmented by the involvement of several pseudopodia together (figure 127 c). In comparison with the axopodia and rhizopodia, the filopodia of the amoebae and Testacea function mainly in the capture of food and less as organelles of locomotion. Incorporation of food particles by cytoplasmic streaming also occurs in the Craspedomonadidae, of the Choanoflagellata. Here the flagellum is surrounded by a cytoplasmic collar, the **collare**, to which the food sticks to be carried away into the cell (figure 16 a).

Gullet-shaped cytoplasmic differentiations, which occupy a specific site in the cell are called **cytostomes** or cell mouths. They are present in some flagellates such as in the genus *Chilomastix*, in which a shorter trailing flagellum at the anterior end washes the food into the cytostome (figure 1 a). Electron microscopy has revealed permanent cytostome structures in Sporozoa as well, so that permeation is not, as was originally supposed, the only process which occurs here. In the intraerythrocytic phase of malaria parasites, the permanent cytostome structure has two concentric rings, so that a cytostome pore is present. Invagination of the cell membrane results in the uptake of erythrocytic cytoplasm at this point with the formation of food vacuoles (figure 140, Cy in 2E, 5E, 8E). The cytostome structures of ciliates are particularly distinct, and they show a wide variation in morphology and function. In the **"whirlers"** the cytostome structure begins peripherally with the peristome cavity. If the peristome is narrowed below into a funnel or tube, this section is called the **vestibulum** or **cytopharynx**. The wall of the peristome and vestibulum is part of the cell surface and is ciliated (figure 127 e). The cilia, often strongly developed, are so arranged in the region of the peristome that their beating produces a vortex of water which washes the food particles into the cytostome, where they are caught (figure 127 d); secretion of slime here can contribute to the trapping of the prey. When the cilia adhere together to form membranellae or membranes, the effect of the peristome ciliature is enhanced. The peritrichous ciliates have a particularly highly developed peristome field; the whole expanse of the flattened front end is surrounded by a substantial spiral of cilia; this is an anticlockwise two-row spiral whose outer row extends as an undulating membrane into the vestibulum (figures 87–89). In the chonotrichous ciliates, the sole cell cilia are those present within the spiral funnel (figure 90). In these attached forms, the cilia function solely in producing the vortex for the uptake of food. When food which has been washed in collects at the bottom of the cytostome (figure 127 e), a food vacuole is cut off and travels into the cell, so that the food may be digested (figure 129).

While the ciliates of the "whirler" type phagocytoze relatively small particles of food, the **"trappers"** are hunters which seize large prey and take them up through an elastic cell mouth. *Didinium nasutum* (figure 74 b) provides an example of this. The exposed cytostome structure has a protruding oral cone (figures 74 b, 128 e) which is furnished at the front with **pexicysts**, spiky organelles for attachment. *Didinium nasutum* is specialized for preying on *Paramecium*. When it makes contact with *Paramecium* it holds it fast by means of the pexicysts. This stimulates the *Paramecium* to defend itself by ejecting its trichocysts (figure 137 d). The *Didinium* is able to dodge this resistance, without releasing the prey, through the adoption by the oral cone, at this stage, of a tubular form. Meanwhile, a paralysing and lethal toxin is injected into the *Paramecium* by toxicysts which surround the pexicysts. The mouth opening of *Didinium* subsequently widens, grasps the prey and sucks it in (figure

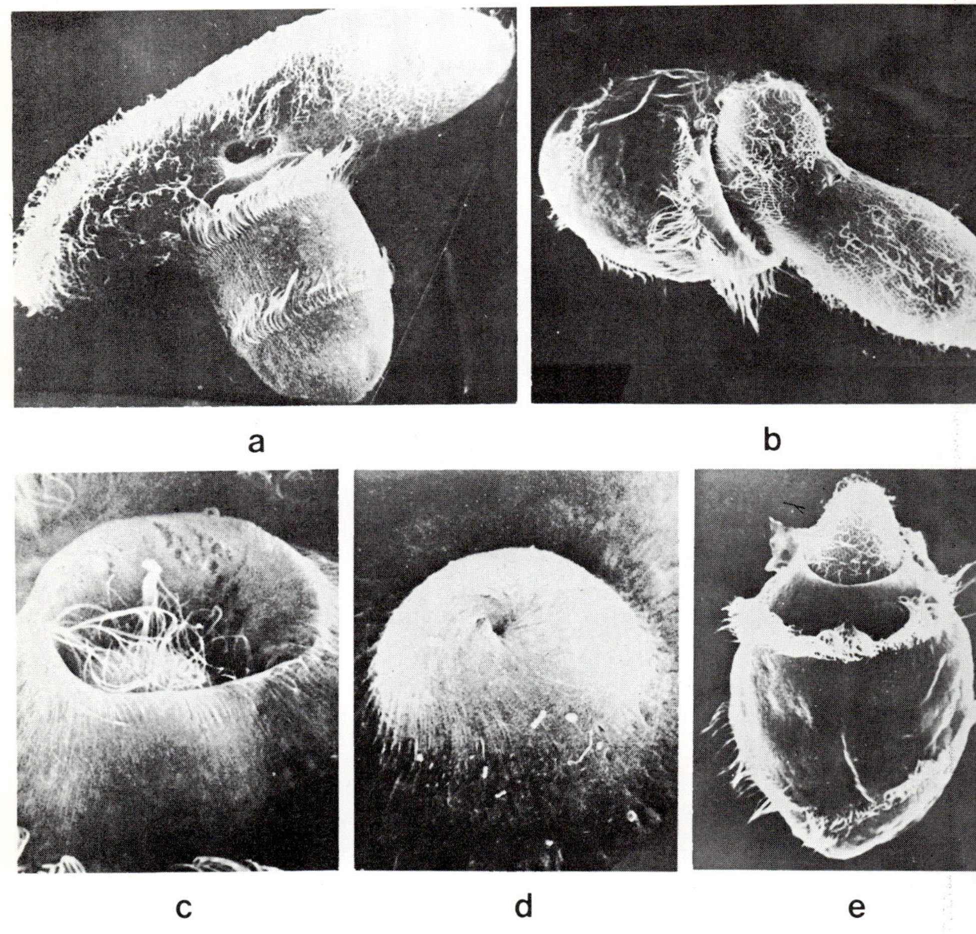

Figure 128. Phagocytosis of *Paramecium multimicronucleatum* by *Didinium nasutum*
   a the cell mouth of *D. nasutum* opens wide and seizes the prey
   b it begins to swallow the prey
   c view into the cell mouth of *D. nasutum* where the prey has almost been swallowed
   d closure of the cell mouth at the peak of the oral cone after phagocytosis has finished
   e *D. nasutum* with the oral cone protruding after the meal. Exposure technique: scanning electron microscope, Magnification a = 430, b = 420, c = 1290, d = 2190, e = 385.

128 a–d). Like *Didinium nasutum*, other "trappers" are also usually specialized to prey on particular species. However, even dead cell matter can also serve as food, for example in *Coleps hirtus* (figure 75 d). The deposition of solidified rods, the **trichites**, can lead to the production of a basket apparatus in the cytostome wall (figure 74 g, h). This stiffening enables the ciliates to nip off algal filaments which serve as food. On the strength of this, Gymnostomata can be differentiated into **Cyrtophorina** with a basket apparatus, and **Rhabdophorina** usually with toxicysts.

The Suctoria can apparently be derived from the predatory Prostomata; they are still holotrichous in ciliature in their immature form as swarmers, but feed in a predatory manner by means of sucking tentacles after they have become attached and are aciliate (figure 91). The tentacles of many Suctoria are used for both trapping and sucking. Haptocysts, short rod-shaped structures with a balloon-like thickening at the end, are situated in the distal enlargement of the tentacle, the tentacle head; when contact is made with the prey, its pellicle is pierced by these haptocysts which hold the ciliate fast so that it can be used for food (figure 91 c). Here, too, a toxic substance can paralyse the ciliate victim. Fibrils are arranged peripherally within the tentacles, and the remaining tube-shaped lumen is filled with cytoplasm. The uptake of the cytoplasm from the prey during the act of sucking proceeds via this lumen. In some Suctoria two types of tentacle are present. For example, *Ephelota gemmipara* (figure 91 m) has long pointed trapping tentacles and shorter sucking tentacles. The trapping tentacles, which can move to the side, also have central fibrils of the microtubular type. After the prey has been captured, it is transferred to the shorter sucking tentacles.

Digestion

The nutrient taken up by the protozoan cell is enclosed by a vesicular constriction of the pellicle, in a food vacuole as soon as it gets in to the cytoplasm. The wall of the vacuole is a unit membrane, whether pinocytosis or phagocytosis has taken place. Enzymatic digestion begins in the food vacuole.

Figure 129 shows the processes of digestion diagrammatically

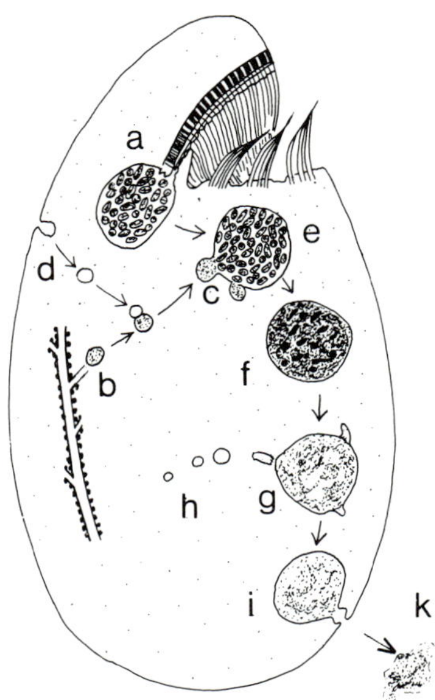

Figure 129. *Tetrahymena pyriformis*. **Diagram of digestion**
a freshly formed food vacuole
b lysosome in the region of the rough endoplasmic reticulum
c lysosomes before penetration into the food vacuole
d formation of pinocytosis vesicles and their fusion with the lysosome
e food vacuole during the uptake of enzymes
f digestive phase of the food vacuole
g diffusion of nutrients derived from the food into the cytoplasm with the production of vacuoles
h the nutrients being carried away in cytoplasmic vacuoles
i expulsion of the undigested food remains (defaecation)
k ball of expelled excrement

using the ciliate *Tetrahymena pyriformis* as an example. After a certain amount of food has been washed in, a food vacuole forms at the end of the cytostome (figure 129 a). In order to set digestion in motion, digestive enzymes must reach the food vacuole. These enzymes are hydrolases which split macromolecules up into fragments of low molecular weight. In ciliates these enzymes are produced in the region of the rough endoplasmic reticulum, and in other cases in association with the Golgi apparatus. The enzymes first appear as "granules" in the form of small vesicles enclosed in a membrane; these are called

lysosomes. Figure 129 b shows a lysosome vesicle still in the region of the rough endoplasmic reticulum. The lysosomes are about 20–40 nm. In the electron microscope they show no characteristic internal structures. They can be concentrated from a cytoplasmic suspension in an ultracentrifuge, according to their specific gravity, in a density gradient. They chiefly contain hydrolases such as α-glucosidase, acid phosphatase, desoxyribonuclease, ribonuclease, protease, α-amylase and β-N-acetylglucosaminidase. Acid phosphatase is the main enzyme of the lysosome.

The production of a food vacuole induces rapid migration of lysosomes to the food vacuole (c), during which the lysosomes to some extent reabsorb the liquid taken in by pinocytosis (d). Only when the enzymes from the lysosome have invaded the food vacuole can the breakdown of the food begin (f). The products of breakdown which are valuable to the cell, the actual nutrients used for structural and energy metabolism, diffuse through the membrane of the digestive vacuole and form small cytoplasmic nutrient vacuoles in the surrounding area of cytoplasm; these then migrate to the site where they will be used (g, h). The vacuole, previously a food vacuole and a digestive vacuole, thus becomes a defaecation vacuole (i), which delivers the indigestible remains of the food to the outside of the cell as balls of excrement (k).

The path of the food vacuoles is not closely defined in the simply organized Protozoa such as the amoebae; it can merely be said that the food vacuoles always occur in the endoplasm. In the more highly organized ciliates, the food vacuole follows a quite definite path, the **cyclosis**. Figure 130 a demonstrates such a migration route with the example of the peritrichous ciliate *Carchesium polypinum*. The food collected at the bottom of the vestibulum (figure 127 e) leads to the detachment of the food vacuole whose cyclosis runs according to the arrowed line. In the lower dotted section, the food vacuole temporarily shows an acid reaction with a pH value of around 4·5. The defaecation of the undigested refuse occurs anteriorly into the vestibulum. The food vacuole of *Paramecium caudatum* has an oval circulation route, shown as a hatched area in figure 130 b. However, depending upon the type of food taken up, a large cyclosis which runs through the whole cell can annex this small cyclosis. The formation and path of the food vacuoles can easily be seen by

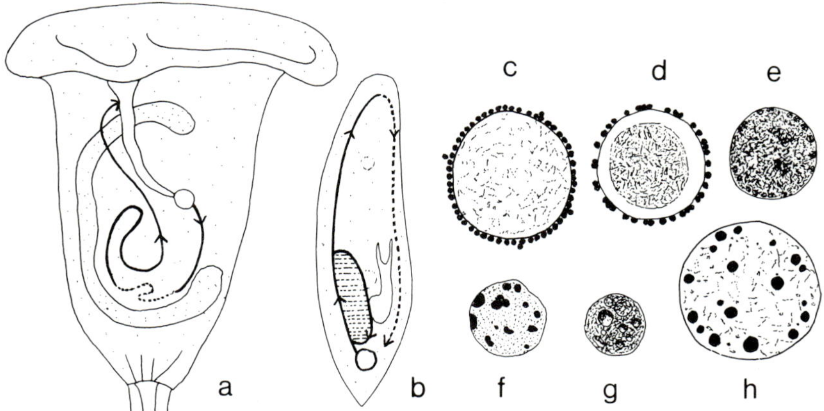

Figure 130. **Paths and alterations of the food vacuoles**
a migration (cyclosis) of the food vacuole in the peritrichous ciliate *Carchesium polypinum* from the base of the vestibulum round to defaecation into the vestibulum. Dotted section of the cyclosis where food vacuole gives an acid reaction
b cyclosis of the food vacuole of *Paramecium caudatum*. Hatched zone = small cyclosis, peripheral route = large cyclosis
c–h changes in the food vacuole of *Paramecium* during cyclosis
    c young food vacuole with enzyme carriers (lysosomes) accumulated on it
    d–g shrinkage of the food vacuole through output of water and conversion to an acid reaction (neutral red staining dark red) with increasing digestion after the penetration of the lysosomes
    h expansion of the vacuole through uptake of water with conversion to the alkaline reaction (neutral red staining yellow)

feeding with carmine granules. Vital staining with neutral red shows a change in the pH value of the vacuoles in *Paramecium* too. Because of extraction of water, the vacuoles become smaller and show an increasingly acid reaction with a pH value of about 4·0 (figure 130 c–g); in this case the enzyme carriers, which are first detected on the periphery of the vacuoles (c, d), penetrate into the vacuole. The food vacuole then grows larger again by the uptake of water, and its contents change over to an alkaline reaction (figure 130 h). The neutral red staining causes the acid vacuoles to appear red; the alkaline, yellow. This alternation of pH values corresponds to the conditions in the digestive tract of higher animals and is necessary, since the effective optimum of

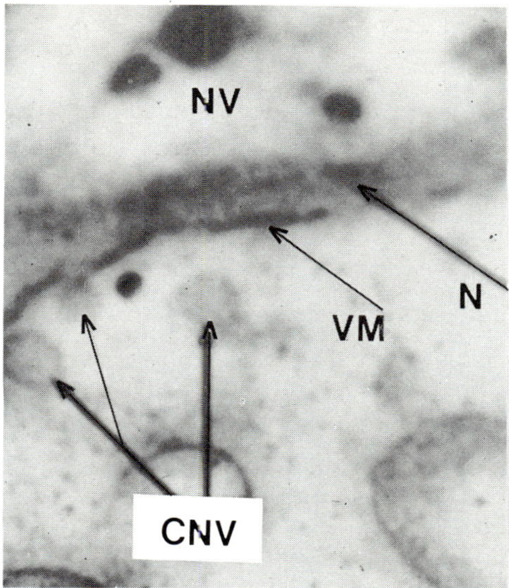

Figure 131. Electron-microscopic exposure. Border of a **food vacuole** (NV) of *Pelomyxa carolinensis*
VM = vacuolar membrane
N = nutrients produced by digestion, within the vacuolar membrane
CNV = cytoplasmic nutrient vacuoles after the diffusion of the nutrients through the membrane

the various digestive enzymes requires different acid or alkaline conditions.

Figure 131 shows an electron-microscopic view of the amoeba *Pelomyxa carolinensis* which shows the transfer of breakdown products obtained from the food by enzyme reactions; these products represent the actual nutrients used for cell metabolism. The amoeba has been fed on *Paramecium* and *Tetrahymena*. The food vacuole, surrounded by the vacuolar membrane, shows a largely homogeneous zone, sited peripherally within the membrane, containing the nutrients which have already been split up and which must reach the cytoplasm of the amoeba for further metabolism. This proceeds at first by means of diffusion through the vacuolar membrane; after this, small nutrient vacuoles of the cytoplasm form on the outer side of the food vacuole (figure 131 CNV). These carry the nutrients to the other

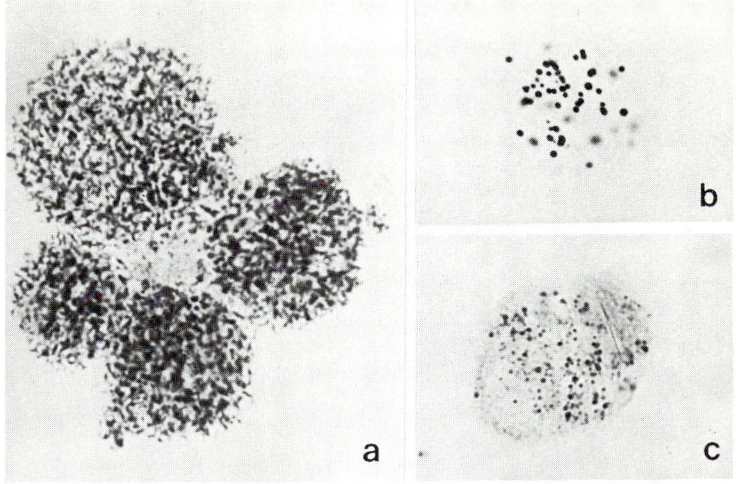

Figure 132. **Enzyme reactions** of *Entamoeba histolytica* in bacteria-free culture with *Crithidia* as food
  a acid phosphatase reaction (after Burstone) in 4 amoebae
  b 1 amoeba after test for nonspecific esterases (after Burstone)
  c test for the enzyme lipoyldehydrogenase (after Scarpelli, Hess and Pearce)

regions of the cell. These cytoplasmic nutrient vacuoles differ from pinocytotic vacuoles in their small size and in their lack of a unit membrane.

Like the digestive processes, all subsequent processes of structural and energy metabolism are also dependent on the participation of specific enzymes. These can be identified by biochemical methods in extracts from concentrated Protozoa, but some can also be demonstrated in the protozoan cell itself by means of cytochemical reactions. In these, a specific substrate added to the cell is chemically altered by the catalytic ability of the enzyme to such an extent that it either becomes a pigment itself or else it unites with an admixed pigment. Figure 132 shows the appearance of *Entamoeba histolytica* in a bacteria-free culture with *Crithidia* provided as food. The figure represents the reaction on acid phosphatase (a), on non-specific esterases (b), and on lipoyldehydrogenase (NAD-dependent dehydrogenase) (c), in which the acid phosphatase is here the most abundant.

The enzymes important for digestion are, as has already been described, prepared in the lysosomes already in the cytoplasm, which is protected from the action of the enzymes by the limiting membrane of the lysosomes. Experimental damage of this membrane by detergents, bile, salts, and alterations of the permeability of the membrane lead to the release of the enzymes and hence to plasmolysis. Under these unphysiological conditions, the lysosomal enzymes lead to the cell's autodestruction.

It has already been seen from figure 129 k that the undigested remains in the food vacuole are discarded by evacuation to the outside. In the Rhizopoda, defaecation occurs at arbitrary sites on the cell surface. Figure 133 shows this process in a small

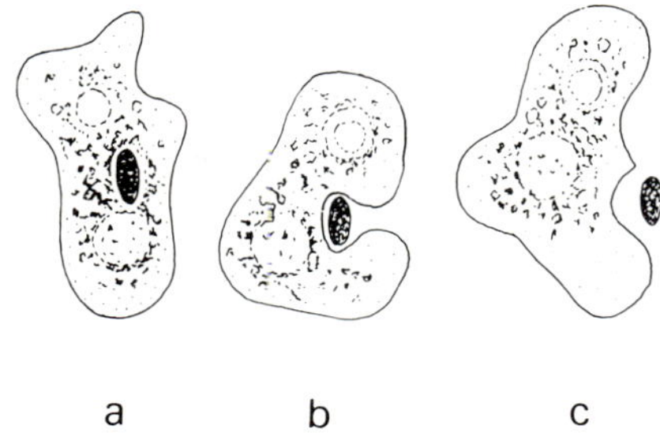

Figure 133. **Defaecation** in an amoeba at an arbitrary site.

free-living amoeba. The vacuole, which has migrated to the pellicle, bursts and empties out its contents. In the ciliates, which have a fixed shape, there is a definite site for defaecation called the cell anus or **cytopyge**. The cytopyge either only appears during defaecation or else, when the pellicle is rigid, can be seen clearly in the interim as well, for example at the back of *Balantidium coli* (figure 1 c), *Eudiplodinium maggii* (figure 83 b) and *Ostracodinium dentatum* (figure 83 d). In some rumen infusoria the cytopyge can extend far into the interior of the organism as a cylindrical canal strengthened by fibrils; carbohydrate deposits make it particularly easy to see (figure 134 b). However, defaecation does not invariably occur in the posterior part of the cell. As can be seen from figure 130 a, it takes place in *Carchesium* into the vestibulum. In the Craspedomonadina, of the Choanoflagellata (figure 16), it occurs within the collar structure.

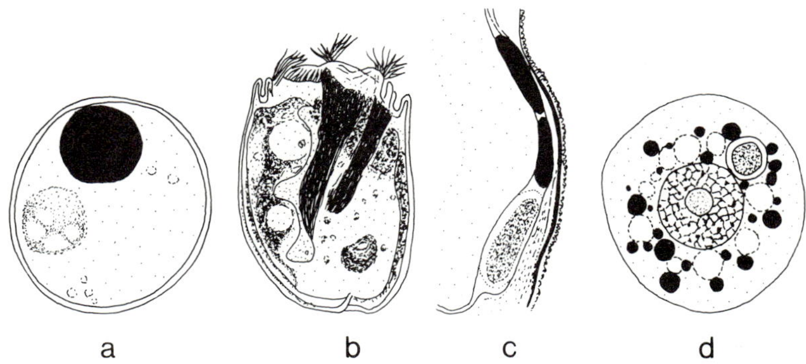

Figure 134. **Reserve substances**
  a cyst of *Iodamoeba bütschlii* with intensively stained glycogen vacuole
  b ciliate *Eudiplodinium medium*, glycogen staining 16 hours after feeding on starch, with deposits formed preferentially in the two elongated skeletal plates, the anal tube and in further fibrillar structures
  c the same animal in transverse section; *above right*, the two skeletal plates rich in amylopectin; *below*, the macronucleus, enclosed by fibrils
  d cytoplasmic region of the Phytomonadine *Haematococcus pluvialis* with numerous darkly pigmented volutin inclusions; above and to the right of the central nucleus, a pyrenoid

Reserve substances

When food supplies are abundant, many Protozoa are able to store up reserve substances in their cells which can be used up when the food supply decreases. This is not important for the blood forms of the African trypanosomes which live in a medium with an excessive supply of glucose; it is therefore hardly surprising that they do not synthesize glycogen. If they are transferred to a glucose-free medium, they die within a few minutes.

The ability to synthesize glycogen is vital for Protozoa which have to live under adverse conditions: hence the promastigote form of *Leishmania donovani* has glycogen sphericles, and the cysts of *Iodamoeba bütschlii* and *E. hystolytica* have glycogen vacuoles (figure 134a). The following energy-storing polysaccharides, which are all polymers of glucose, have been found in Protozoa: glycogen in Tetrahymena, Trichomonas, Prototheca and the above-mentioned amoebae and Gregarinida; starch in

Polytomella, Polytoma and Chilomonas; amylopectin in the rumen infusoria and *Eimeria*; cellulose in Acanthamoeba; and a glycogen polymer with β-1-3-linkages in Ochromonas, Astasia and Euglena (= paramylum). In addition there is leucosin in Chrysomonadina and Heterochloridina, and oils and fats in Chrysomonadina, Cryptomonadina, dinoflagellates and Chloromonadina. Sufficient reserve substances are necessary, particularly for the metabolism of cysts which, after all, no longer take up food for themselves. Thus the cyst of *Entamoeba histolytica* produces not only a glycogen vacuole but also chromatoid bodies which stain intensively with nuclear strains (figure 116 e, f) and which are used up again during the maturation of the cyst which involves nuclear divisions (figure 116 g). These chromatoid bodies are aggregates of ribonucleoprotein which can already be seen by electron microscopy, during the stage preceding cyst formation, as finely distributed crystalloid-stratified complexes in the cytoplasm of the amoeba.

While the polysaccharide deposits are usually finely distributed in the cytoplasm in the vegetative stages, they can lead to the formation of granules in the Gregarinida. They also tend to occur more at particular places in the cell. Special glycogen vacuoles are produced in the cysts of the intestinal amoebae; they stain intensively with the addition of iodine or other glycogen stains (figure 134 a). If the cyst is expelled with the faeces, the glycogen vacuole, like the chromatoid bodies of *E. histolytica*, is slowly used up. In the rumen infusoria, abundant amylopectin is deposited at certain places on fibrillar **structures. In some genera there are attenuated fine-meshed honeycomb-like nets of fibrils, the so-called skeletal plates,** in addition to the universal peripheral fibrillar systems (figure 117 b). All these fibrillar systems are places where storage tends to take place. Figure 134 b shows an example of this in a whole preparation. The cylindrical anal tube above the cytopyge also has intensive deposits of carbohydrates. Figure 134 c shows a transverse section through this ciliate. Above right are the two skeletal plates containing reserve substances; below is the macronucleus which has been cut through. After it has been fed with an abundance of starch, the whole cell gives an intense polysaccharide staining reaction. Figure 134 b and c show the condition of the cell 16 hours later. 50–60 hours after feeding, all the amylopectin contained in the cell has been used up.

Fats and oils in droplet form also frequently occur as reserve substances in the heterotrophic Protozoa; these are mainly produced by carbohydrate metabolism, and occur in particularly large amounts in Radiolaria, in the dinoflagellate *Noctiluca miliaris*, and in *Eimeria gadi*, a coccidian occurring in the air bladder of the cod. Excessive accumulation of fats can lead to pathological conditions in Protozoa, just as in Metazoa; fatty degeneration has been observed, for example, in *Polytoma uvella* in the Phytomonadina. In some Protozoa, excluding the ciliates, staining reveals inclusions which have a strong affinity for basic stains; in this case, a change of colour, a metachromasis, can occur. Figure 134d shows inclusions of this sort in *Haematococcus pluvialis* in the Phytomonadina. Treatment with ribonuclease has shown that compounds containing RNA are involved here. The inclusions are rapidly used up in culture media lacking phosphorus, and the flagellates encyst or perish. These metachromatic deposits are called **volutin**. They are stores of phosphate, contain long-chain polyphosphates, and are important in cell division.

Osmoregulation and excretion

The cytoplasm, with its organic substances and mineral salts, has a higher osmotic pressure than the environment in the case of fresh-water Protozoa. Because of the permeability of the pellicle, water constantly diffuses from this hypotonic medium into the cytoplasm. In order to counteract this osmotic gradient, such Protozoa have a pump system in the form of the contractile vacuoles, through which the fluid which has been taken up is returned to the outside, so that physiological pressure is maintained. The most striking are the contractile vacuoles of some ciliates. Figure 77a shows (left) a vacuole of this sort in *Frontonia leucas* with long nephridial canals which conduct the fluid to the contractile vacuole. *Paramecium caudatum* (figure 77b) has two contractile vacuoles, which function in an alternate rhythm. In the diastolic phase the central reservoir, the actual vacuole, is filled (figure 135b). A sudden contraction empties out the contents of the vacuole into the environment. This phase is called **systole** (figure 135a). The nephridial canals, now dilating, conduct more fluid to the central vacuolar vesicle, so

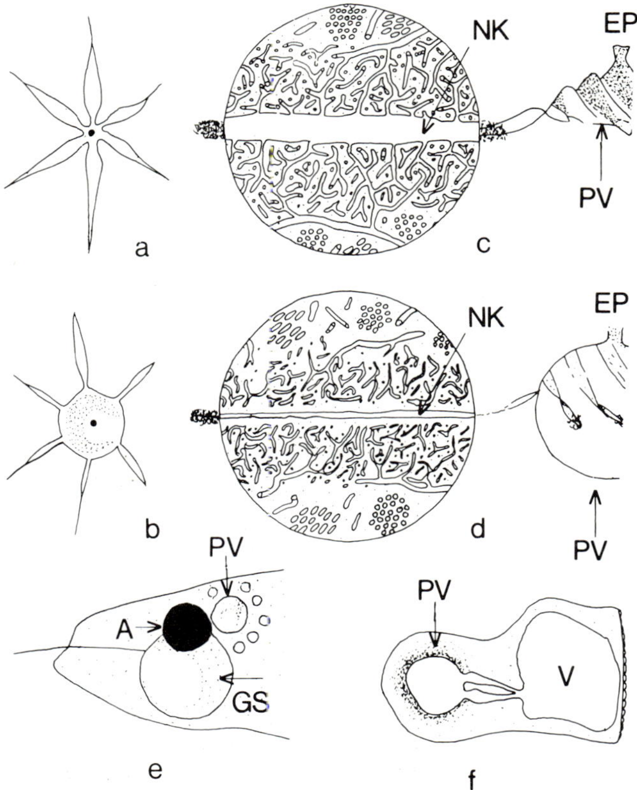

Figure 135. **Contractile vacuoles**

a–b the contractile vacuole of *Paramecium caudatum* seen under the light microscope

  a nephridial canals in diastole, vacuole (central reservoir) in systole

  b nephridial canals in systole, vacuole in diastole

c–d *Paramecium aurelia*, electron-microscopic appearance of a nephridial canal (NK) with part of the contractile vacuole (PV) with myonemes and excretory pore (EP) and the conducting ampulla

  c nephridial canal in diastole in open contact with the conducting nephridial tubuli, ampulla in diastole, contractile vacuole in systole

  d nephridial canal in systole without contact with the nephridial tubuli, ampulla in systole, contractile vacuole in diastole

  e *Euglena viridis*, contractile vacuole (PV) surrounded by the conducting secondary vacuoles, A = eyespot (stigma), GS = flagellar sac

  f *Campanella umbellaria*, contractile vacuole (PV) with 2 excretory canals discharging into the vestibulum (V)

that diastole of the central vacuole begins again. An analysis of this process is possible by electron microscopy. In figure 135 c the nephridial canal (NK) is in diastole, and the central contractile vacuole (PV) in systole. The entry of the fluid from the cytoplasm into the nephridial canal proceeds by a system of reticularly-branched nephridial tubules which have an open connection with the nephridial canal on one side, and are in contact with the endoplasmic reticulum on the other side. The nephridial canal is linked with the contractile vacuole by the ampulla via the constricted injection canal. Contraction of the nephridial canal causes the fluid to pass on along this route to the central contractile vacuole; as this happens, contact between the nephridial canal and the network of nephridial tubules is broken to prevent the fluid flowing back into the cytoplasm. Figure 135 d shows this phase with the nephridial canal in systole and the contractile vacuole in diastole. In the subsequent systole of the vacuole, contractile fibrils of the vacuolar vesicle cause a contraction of the ampulla and its injection canal at the same time, so that the fluid which has collected is passed out through the pellicle into the environment via an excretory pore (EP) which can be closed by a membrane.

The principle of filling the contractile vacuoles through nephridial canals by no means holds for all Protozoa. In the simplest case, simple vesicles, which differ from the normal fluid vacuoles only in that they empty their contents outside the cell, occur in the cytoplasm. In some Protozoa a process occurs which is the reverse of that shown in figure 131 for the distribution of dissolved nutrients emanating from the food vacuole. In *Euglena viridis*, for example (figure 135 e), the fluid is brought by small conducting secondary vacuoles and passed on to the contractile vacuole. The contractile vacuole then empties its contents into the flagellar sac, which is connected to the outside world as shown in figure 139 b and c. This diastolic filling of the contractile vacuole by secondary vacuoles also occurs in many amoebae, in which the vacuole, when full, empties its contents directly to the outside, anywhere on the cell surface, just like the digestive vacuole.

In the flagellates and ciliates which have a constant form, the positions of the contractile vacuole and of the excretory pore are clearly defined. In the peritrichous ciliate *Campanella umbellaria*, the emptying of the contractile vacuole via the excretory canals into the vestibulum (figure 135 f) corresponds to the behaviour of the food vacuole. The number of contractile vacuoles, too, differs according to protozoan species, and the rhythm of contractions, the pulse frequency, is similarly species-specific. The diastolic phase lasts 5–10 seconds in *Paramecium caudatum*, and about 30 seconds in the flagellate *Euglena ehrenbergi* and the ciliate *Coleps hirtus*. *Amoeba proteus* needs about 5 minutes for the production of its 50-μm vacuole, and the marine suctorian *Acineria incurvata* requires 6–12 minutes. Basically, the pulse frequency in marine and parasitic Protozoa is slower than that of fresh-water forms, because of the lower osmotic difference. Species like *Actinophrys sol* and *Thecamoeba verrucosa*, which can become accustomed to sea water instead of fresh water, lose their contractile vacuole as they do so. The pulse frequency is influenced not only by the salt content of the habitat, but also by other environmental factors such as the composition of the medium and especially the temperature. In the hypotrichous ciliates *Stylonychia pustulata* and *Euplotes charon*, their respective pulsation rates change from 18 and 61 seconds at $+5\,^{\circ}\mathrm{C}$, to 4 and 23 seconds at $+30\,^{\circ}\mathrm{C}$.

In addition to the fluids which are eliminated through the contractile vacuole, or directly through the pellicle, and the solid food remains which are emptied from the food vacuole by defaecation, secretions can also arise during metabolism which are deposited in solid form within the cell itself. These are called **secretions** if they are of use to the Protozoa for hardening the coats, shells, capsules, skeletons and stalk structures, but are called **excretions** if no such use can be seen. In *Paramecia* these excretory crystals consist mainly of calcium phosphate. They are distributed throughout the cytoplasm and shine brightly by birefringence when the unstained fresh preparation is looked at in polarized light (figure 136 a). In the ciliate *Stentor coeruleus* the excretory inclusions in the form of a blue-green pigment result in pigmentation of the whole cell. In the malaria parasites a brown pigment, like haematin, forms in the cytoplasm of the parasite derived from the haemoglobin of the blood corpuscles. Figure 136 b shows its granular structure in a macrogametocyte of *Plasmodium berghei*, a parasite of small rodents, in a Giemsa-stained preparation. This pigment formation becomes clearer in an unstained smear seen by incident light using reflected-light microscopy (figure 136 c).

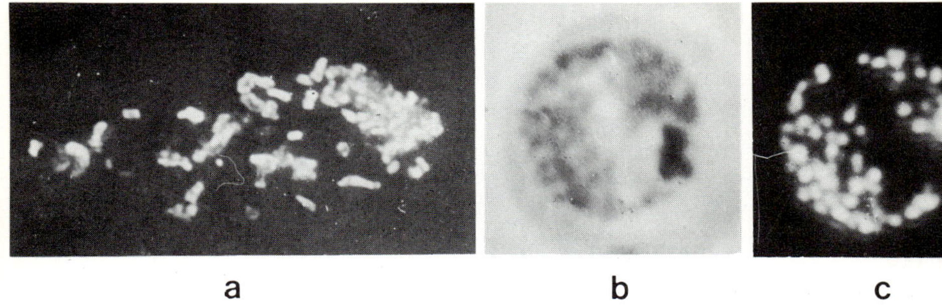

a   b   c

Figure 136. **Excretions**
  a Excretory crystals of an unstained **Paramecium** in polarized light
b–c *Plasmodium berghei*, macrogametocyte in an erythrocyte
  b after Giemsa staining
  c pigment content of the same gamont in unstained preparation under reflected-light microscopy

In some Protozoa the excreted substances can be eliminated by the formation of excretory vacuoles and the discharge of these at the surface of the cell. In ciliates, the excretions perhaps also disintegrate in the region of the contractile vacuoles, whose significance in the excretory process can, however, not yet be satisfactorily evaluated. In the Foraminifera, the pigmented excretions known as **xanthosomes** are cemented with the faecal mass into spherical or ellipsoidal stercomes and eliminated at defaecation. Because of their durability, these stercomes form large deposits in places where there are large numbers of Foraminifera.

Extrusomes

Extrusome is the name given to organelles of various species which are shot out of the organism in response to a stimulus. They include the polar threads of the Cnidosporidia which have already been described; these lie in the polar capsule in the Myxosporidia and Actinomyxidia (figure 66 b) but only in a simple vacuole in the Microsporidia. They are organelles for attachment, which are released when stimulated by the digestive juices in the gut of a new host (figure 69 i). The nematocysts,

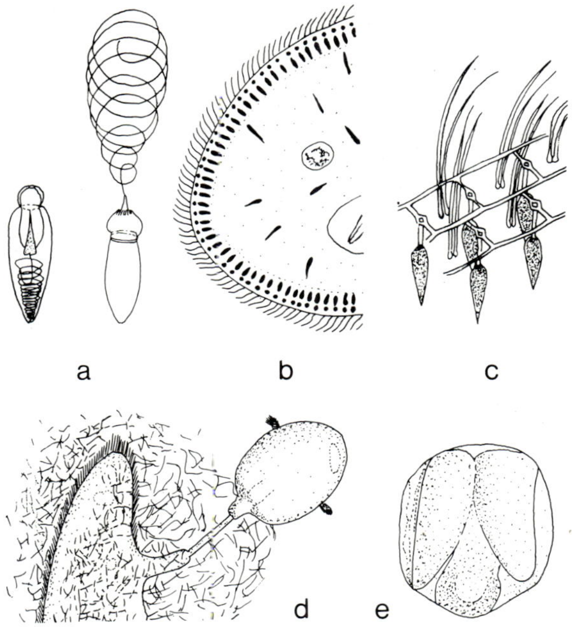

Figure 137. **Extrusomes**
  a Sting capsule of *Polykrikos* sp. (dinoflagellate) with coiled thread and spike
  b section through *Frontonia leucas* with subpellicular trichocysts; inside, developing vacuoles and young trichocysts
  c *Paramecium nephridiatum* with trichocysts on the transverse ridges of the pellicle and 2 cilia in the centre of each enclosed field
  d part of Paramecium with expelled trichocysts after attack by *Didinium nasutum*. *D. nasutum* with the tube of the oral cone thrust out
  e ejectisome of *Chilomonas paramecium* (Cryptomonadina) before ejection

"sting capsules", function as defence organelles; these occur in the phylogenetically interesting dinoflagellates in the genera *Polykrikos* (figure 5 k) and *Nematodinium*. They resemble the sting capsules of the coelenterates, as shown in figure 137a. They arise autogenetically by a special form of self-differentiation. The trichocysts, which are produced in many holotrichous ciliates, are also defence organelles. In the genera *Frontonia* and *Paramecium*, they lie assembled under the cell surface in large quantities (figure 77 a, b). They form inside the cell in cytoplasmic vesicles and migrate to the pellicle as rod-shaped structures (figure 137 b). In *Paramecia* they are anchored to the ridges of the pellicle, whereas the cilia arise from the centres of the fields formed by the ridges (figure 137 c). The trichocysts shoot out through window-like openings in the ridge structure when strong mechanical or chemical stimuli influence the ciliates, e.g. when attacked by *Didinium nasutum* (figure 137 d) or experimentally by dilute acetic acid. While the trichocysts of some holotrichs (*Paramecium*) are completely expelled, those of other species (*Frontonia, Lionotus*) remain with their ends lodged in the pellicle. Figure 138 a shows the electron-microscopic appearance of a subpellicular trichocyst in *Paramecium caudatum*. The ripe trichocyst consists of a large shaft upon which sits a spike which is itself surrounded by a cap. After ejection, the appearance of the spike remains unchanged, while the shaft (which has increased in length about ten-fold) is transversely striated with a periodicity of 55 nm (figure 138 b). Salts are also deposited in this shaft which, however, consists chiefly of filamentous proteins; thus the trichocysts also have a certain osmoregulatory function. This is also suggested by the occasional apparently spontaneous ejection of trichocysts. Trichocysts occur not only in ciliates but also in flagellates, namely in many dinoflagellates.

In the holotrichous predators, the Gymnostomata, extrusomes are used as weapons for attack. When they eject, a paralysing and lethal toxin penetrates into the prey. These extrusomes known as **toxicysts**, which are produced in cytoplasmic vesicles as are the trichocysts, are deposited in the region of the cytostome. They take the form of tube-shaped, often slightly twisted capsules from which a thread shoots out as in the sting capsules (figure 137 a) in a telescopic fashion or by evagination like the finger of a glove; the toxin reaches the prey through this thread. In *Didinium nasutum* the toxicysts surround the pexicysts by which the prey is at first held fast. The *Paramecia* which have been attacked react to this by exploding their trichocysts, which the *Didinia* evade by the adoption of a tubular structure by their oral cone (figure 137 d). Phagocytosis of the prey by the widened mouth opening of *Didinium nasutum* (figure 128) proceeds only after the toxicysts have had their paralysing effect.

The extrusomes deposited in the pellicle of the vestibulum in Cryptomonadina are called **ejectisomes**; these too expel a

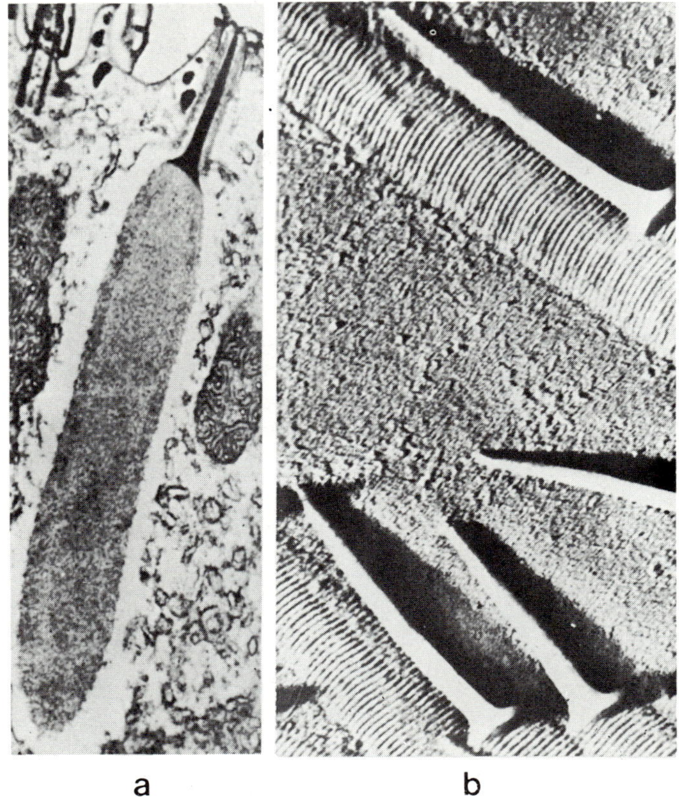

a   b

Figure 138. Electron-microscopic view of the **trichocysts of Paramecium**:
a Subpellicular trichocyst of *P. caudatum* with a spike sitting on a shaft under a cap (magnification 19,900 ×)
b ejected trichocysts, free spikes unchanged, the elongated shafts transversely striate. Exposure after metal shadowing (magnification 15,420 ×)

thread in an explosive manner. The structures are at first in two contiguous parts and are relatively-short externally-tapering cylinders surrounded by a membrane. The cylinders are evidently formed by the tight coiling of the thread which is later released (figures 137 e, 108 E). Finally, subpellicular extrusomes containing slime (mucocysts) occur in some ciliates. Here, too, evacuation represents a reaction to an outside stimulus. These mucocysts have been demonstrated in, for example, the genera *Colpidium* and *Tetrahymena* (figures 77 c,

78 d); their extruded slime allows the ciliate to produce a protective coat.

Sensory organelles and reactions to stimuli

The cytoplasm of the Protozoa can react to environmental influences with behavioural changes; it is, in other words, susceptible to stimuli. This is provided that the stimulatory effect reaches a sufficient threshold value. There are various types of reaction. They include, for example, food uptake, the onset of sexual processes, changes of shape such as contractions and, in many cases, movements in response to stimuli which are called **taxes**. If the induced movement is towards the stimulus, the taxis is positive and, if away from the stimulus, the taxis is negative. A direct alignment towards or away from the stimulus is **topotaxis**. An escape movement which is merely evasive is **phobotaxis**. The responsive movements are named according to the type of stimulus operating: **thermotaxis** is response to temperature, **chemotaxis** to chemicals, **phototaxis** to light, **galvanotaxis** to electric current and voltage, and **barotaxis** to pressure stimuli in general, comprising **thigmotaxis** to contact, **rheotaxis** to water current, and **geotaxis** to the earth's gravitational pull.

The amoebae, with few defined organelles, are able to react; this shows the capacity of the cytoplasm to respond to stimuli, even when no specific cytoplasmic structures are present. The stimulus may be detected by the whole protozoan cell, or certain parts of the cell may prove to be particularly sensitive to stimuli. When living paramecia are cut up, the individual pieces remain sensitive to contact; i.e. they continue to be thigmotactic. If the cut is just in front of the cell mouth, the posterior part continues to be capable of chemotaxis and thigmotaxis, but not of thermotaxis. The sensitivity to temperature must therefore be localized in the anterior part of the cell. If the cut passes through the cell mouth, chemotaxis is also lost. If this sort of localization of response to stimuli is linked to specific cell differentiations, these are called **sense organelles**. In the flagellate amoeboid forms, the Rhizomastigina, the flagella function as tactile organelles, not as organelles for locomotion. The tactile function is also attributed to the flagella of other flagellates. The cilia of the ciliates are also sensitive to pressure and current, particularly certain stiffer individual cilia, e.g. on the dorsal surface of Hypotricha or in the caudal tuft of *Paramecium caudatum.* They help to stabilize locomotion. In this process, *Paramecium* is positively rheotactic in reaction, as are many other ciliates, and hence swims against the water current. The trailing flagella of some flagellates also show rheotactic sensitivity. But some cilia can detect not only mechanical stimuli but also chemical stimuli. In studies of food uptake in Entodiniomorpha, many animals when feeding will at first wash granules of charcoal or carmine in by means of their membranellae, just as they do grains of rice starch. Soon, however, only rice starch is taken up. Powdered chalk is never taken up; indeed, the cilia are often suddenly retracted.

Positive **phototaxis** in the Protozoa occurs particularly in the pigmented phytoflagellates, whose metabolism requires light. For this reason, some of them produce specific photo-sensory organelles, the **stigma** or **eyespots**. Some marine dino-flagellates form highly developed eyespots. Stratified hyaline bodies which appear homogeneous by light microscopy but which by electron microscopy show differentiation, form a lens system which effects an intensification of the light incident on a light-sensitive absorbing pigment (figure 139 a). In this, light energy is converted to chemical energy. In simple cases, for example in *Euglena*, a pigmented eyespot, a stigma, serves merely to cast a shadow in a region of light-sensitive cytoplasm, so that the direction of the incident light is ascertained (figure 139 b). A red spot is also situated in front of the light-sensitive pigment zone in figure 139 a. The stigmata are often differentiations of the chloroplasts; however, in the Euglenoidina they lie laterally at the bottom of the flagellar sac. As in some Astasia, species of Euglenoidina which no longer have plastids may also lack a stigma (figure 139 c). While the autotrophic flagellates are positively phototactile, most heterotrophic Rhizopoda and ciliates show negative phototaxis, provided they do not contain autotrophic symbionts, the zoochlorellae, as found in *Paramecium bursaria.*

Chemotaxis has great significance. Dilute fatty acids, e.g. butyric acid in the proportion $1:10^6$, have a strongly positive effect on some flagellates (*Euglena, Astasia*). The pH value has been found to be important in the chemotaxis of paramecia,

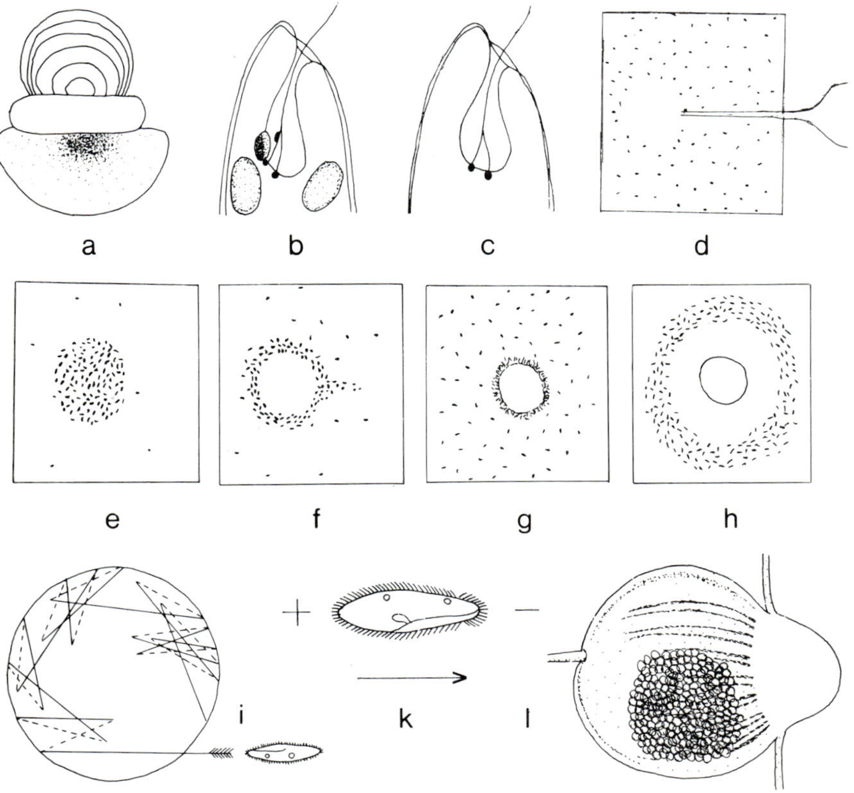

Figure 139. **Sensitive organelles and reactions to stimuli.**
a eye structure in the dinoflagellate *Erythropsis richardi*. Below: pigment cup with red spot (dark zone); above: homogeneous inner and stratified outer hyaline bodies = lens structure
b *Euglena* sp. with chromatophores, with stigma below and to the left of the flagellar sac
c *Astasia* sp., euglenoid without chromatophores and without eyespot
d–h Effect of chemical stimuli on cultures of *Paramecium aurelia*
d evasive reaction after injection of a negatively chemotactic fluid
e collection in a positively chemotactic drop
f as in e, but with too high a concentration: collection in region of optimal dilution
g gathering round a positively chemotactic bubble of carbon dioxide
h the same preparation as in g after diffusion has taken place
i swimming movement of *Paramecium* within a drop with positive chemotaxis: a trap effect
k galvanotaxis in *Paramecium* in the direction of the cathode

with an optimum for *P. caudatum* at pH 5·4–6·4. If a drop evoking a negative chemotactic response is added to a uniform suspension of living paramecia, the paramecia draw away from this place straight away (figure 139 d). Conversely, they collect together very quickly in a chemotactically positive drop (e). If the latter solution is present at too high a concentration, however, it has a chemotactically negative effect, so that the paramecia accumulate in the region of optimal diffusion (f). A bubble of carbon dioxide has a chemotactically positive action (g). After increasing diffusion, the zone of optimal concentration is sought out in this case too (h). If the paramecia have swum into a chemotactically positive drop as in figure 139 e, this has a trap effect. A phobotactic evasion motion occurs now and then at the outer limit of the drop of fluid, with an aimless dodging movement and return, so that this process is repeated often (figure 139 i). Chemotaxis is very important in the selection of food, for location of a sexual partner and, in parasitology, in the form of organotropy. The malarial sporozoites can be taken as examples of this; after the oocysts have burst, they circulate with the haemolymph of the mosquitoes throughout the whole body and are finally gathered, by the trap effect, in the salivary glands. After transmission to the blood of the vertebrate host, the causal agents of malaria in man collect only in the parenchymatous cells of the liver. Only the later merozoites have an organotropy towards the erythrocytes. Chemotaxis thus varies between the individual developmental stages of the same parasite.

Thermotaxis, too, proceeds mainly as phobotaxis, as in the hypotrichous ciliate Oxytricha. In this case, too, there are species-specific optimal ranges which, for example, lie between 24 and 28°C for *Paramecium*. The reaction of the Protozoa to electro-stimuli is also species-specific. *Trachelomonas*, amoebae and many ciliates normally move to the cathode (figure 139 k). *Spirostomum* adopts a transverse position across the current. *Polytoma, Chilomonas* and *Opalina* migrate to the anode. As is the case for many stimuli, intensity of stimulus also has an effect. Whilst *Paramecium* migrates to the cathode when the current is weak, a stronger current activates the cilia at the posterior end,

---

1 statocyst of the parasitic holotrichous ciliate **Blepharoprosthium**; *right*, pellicular cap structure; inside, mass of granules; *left*, emerging centripetal fibril.

which beat in opposition to the others, thereby causing a return to the anode.

Geotaxis. Some ciliates have special statocysts (figure 139l) for detecting the earth's gravitational pull; these are vesicles in which enclosed granules exert pressure on whichever is the lower side at the time. These also include the conspicuous excretory vacuoles **(Müller corpuscles)** of the genera *Remanella* (figure 74f) and *Loxodes* (figure 75f). In the rest of the Protozoa, the inclusions of the normal excretory and food vacuoles distributed in the cytoplasm assume the role of geotactic indicators. Most ciliates show a negatively geotactic response. Thus *Paramecium* rises to the surface. After phagocytosis of finely powdered iron, the pressure effect in the food vacuoles of paramecia can, however, be altered in a magnetic field, resulting in a motion along the lines of magnetic force.

This example of geotaxis in the ciliates also indicates a conduction of the stimulus from the sensory sites, the vacuoles, to the responsive organelles, the cilia, through the cytoplasm itself without special stimulus-conducting organelles. If paramecia are constricted in the middle with a glass thread, the beating of the cilia on the anterior end becomes independent of that of the cilia of the posterior end as soon as the internal cytoplasmic column is interrupted. On the other hand, if the surface of a ciliate is severed, the organized beat of the cilia is interrupted, from which it can be concluded that the beat is co-ordinated among the cilia by the subpellicular fibrillar system.

Functional morphology

A consideration of the various cell organelles reveals that the functions of a protozoan cell are linked at times to certain structural elements and forms. These are, however, by no means always constant, but can change in a temporal sequence. In the Suctoria, only the swarmers are ciliated in order to seek out a new habitat, while the non-ciliated adult stages develop a stalk in order to attach themselves in this habitat, and they develop tentacles in order to trap and consume their prey. Each stage of development has a morphology and function suited to the needs of the moment. This changeability is clear even with light microscopy, particularly in Protozoa with a distinct life cycle, e.g. in the life cycle of the malarial parasite in the production of slender motile sporozoites, whose duty it is to penetrate into new host cells, in contrast to the subsequent much-larger rounded schizonts and gamonts (figures 60, 140). In this example, the morphological and functional differences between the motile microgametes and the nutrient-rich larger macrogametes is appropriate. This change in form, which is dependent on function, can be recognized in the fine structure seen in the electron microscope, as well as in the structures which can be discerned under the light microscope.

The numerous membrane systems which can be seen in a protozoan cell are not static structures. Rather, they are subject to constant variations. Their condition is dependent upon the physiological state of the organism and can be influenced by external factors. In the amoeba *Pelomyxa illinoisensis*, profound changes in the mitochondrial tubules appear during mitosis. Differences in structure, position and number of the mitochondria as a function of age are known in the ciliate *Tetrahymena pyriformis*. When *Plasmodium fallax* is exposed to the action of 8-aminoquinolines, the mitochondria become distended. Membranes are, however, not only capable of structural changes but can, as is typical for the Golgi apparatus, cut off vesicles, undergo complete reversion, and be rebuilt as part of the ground cytoplasm. The dynamics of these processes are therefore aptly called "membrane flow". In Protozoa, structures necessary for a particular function can arise very quickly. In the transition from the amoeboid to the flagellate state, the flagella of *Naegleria gruberi* grow at a rate of 0·5 μm in length per minute.

In the Protozoa with a cyclical development there is a regular change also in structures seen by electron microscopy. The developmental cycle of the *Plasmodium* spp. is characterized by an alternation of hosts, involving insects on the one hand and various vertebrates on the other. The individual stages develop intracellularly in various tissues. Electron-microscopic studies of the sporozoites and the merozoites of the tissue phase and blood phase have shown that structures are produced for particular functions, and regress when they are no longer required. In this case, electron microscopy has made an important contribution to the understanding of functional morphology.

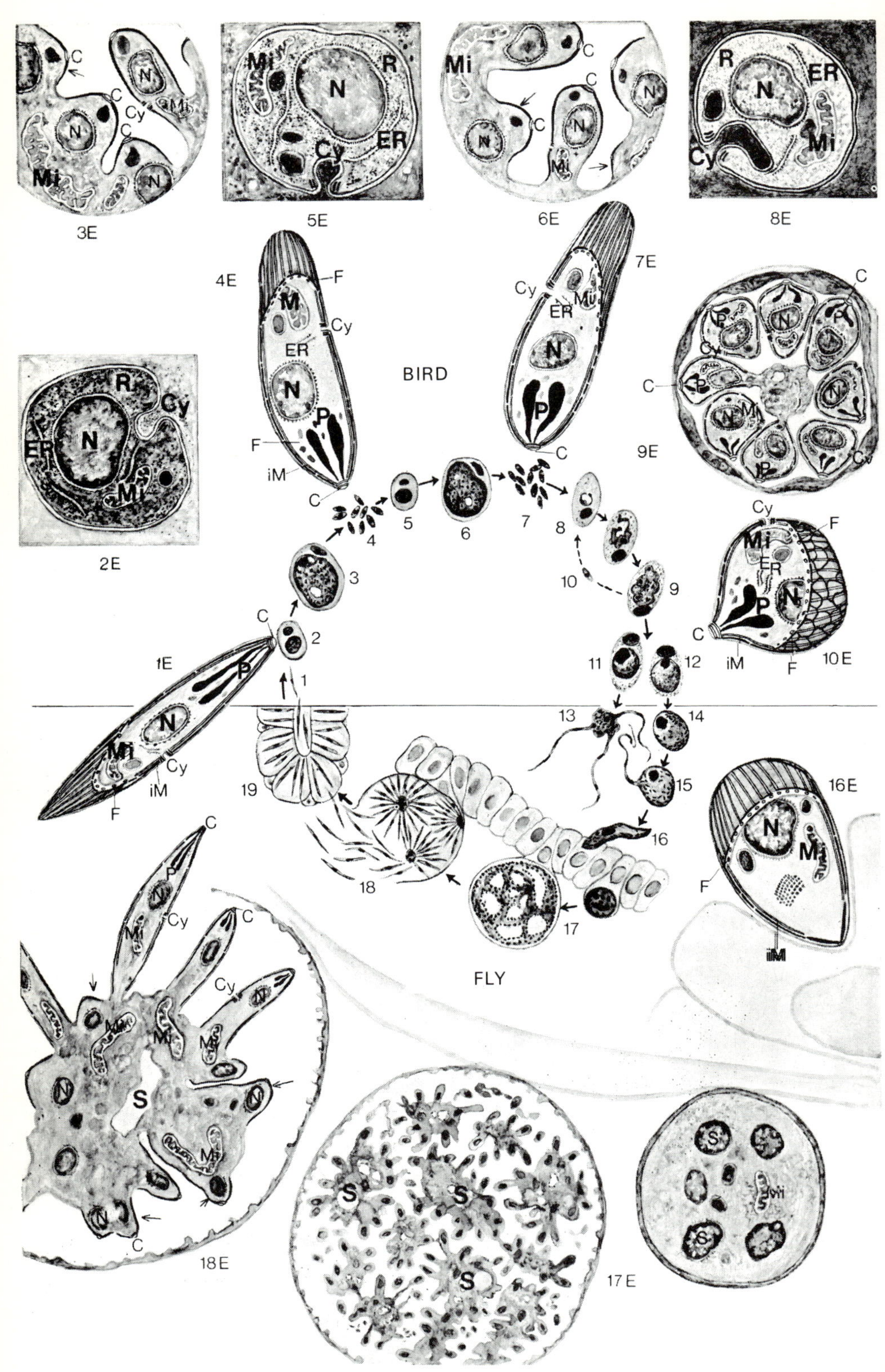

3E
5E
6E
8E
C
N
C
Cy
Mi
C
N
C
N
Mi
N
Mi W
N
R
Cy
ER
Mi
C
C
C
N
N
Mi
R
N
ER
N
Cy
Mi
2E
4E
BIRD
7E
9E
R
Cy
ER
N
Mi
F
M
Cy
ER
N
P
F
iM
C
Cy
Mi
ER
N
P
C
C
N
P
Cy
P
N
Mi
P
C
Cy
C
3
4
5
6
7
8
10
9
C
2
11
12
Cy
Mi
ER
P
N
C
iM
F
F
1E
1
13
14
10E
P
15
16E
N
C
N
16
N
Mi
Mi
Cy
iM
F
19
18
17
F
iM
FLY
C
P
N
Mi
Cy
C
Cy
C
N
Mi
Mi
Mi
N
S
N
C
18E
S
S
S
S
S
S
Mi
17E

Sporozoites and merozoites are stages which must penetrate into cells in order to develop further. Before this penetration, special organelles and structures are added to the anterior region of the cell in these stages of development. These include: apical structures, comparable to the conoid of other Sporozoa, provided with concentric rings; the paired organelle; an additional inner cell membrane and subpellicular fibrils—structures which belong to the apical complex (see page 54). Since these structures disperse immediately after penetration into the cell, it is assumed that they are required to overcome the resistance of the host cell membrane. In this case, the apical structures would be responsible, in conjunction with the paired

Figure 140. **Simplified developmental cycle of an avian plasmodium** (compare figure 60, cycle of *Plasmodium vivax*)
1–19: appearance under light microscopy
1E–18E: appearance under electron microscopy
   1, 1E = sporozoites which invade the cells of the reticulo-endothelial system of the bird within 30 minutes of the insect bite (2, 2E). Pre-erythrocytic cycle (2–7) within these cells
   3, 3E = schizogony with formation of merozoites (cryptozoites 4, 4E). Repeated schizogony (5, 5E, 6, 6E) with production of further merozoites (metacryptozoites 7, 7E)
8–12 = erythrocytic developmental phase
   9, 9E = schizogony
   10, 10E = merozoite
   11 = microgamont (microgametocyte)
   12 = macrogamont (macrogametocyte)
13–15 = gametogamy
   13 = production of microgametes (with flagella)
   15 = fertilization
   16, 16E = ookinete (zygote)
   17, 17E, 18, 18E = development of oocyst in the gut wall of the fly: production of sporozoites, which migrate into the salivary gland via the coelome and haemolymph after rupture of the oocyst (18)
C = conoid-like structure
Cy = cytostome
ER = endoplasmic reticulum
F = subpellicular fibrils
iM = inner cell membrane
Mi = mitochondria
N = cell nucleus
P = paired organelle
R = ribosomes
S = sporoblast
← = initial thickening of the cell membrane

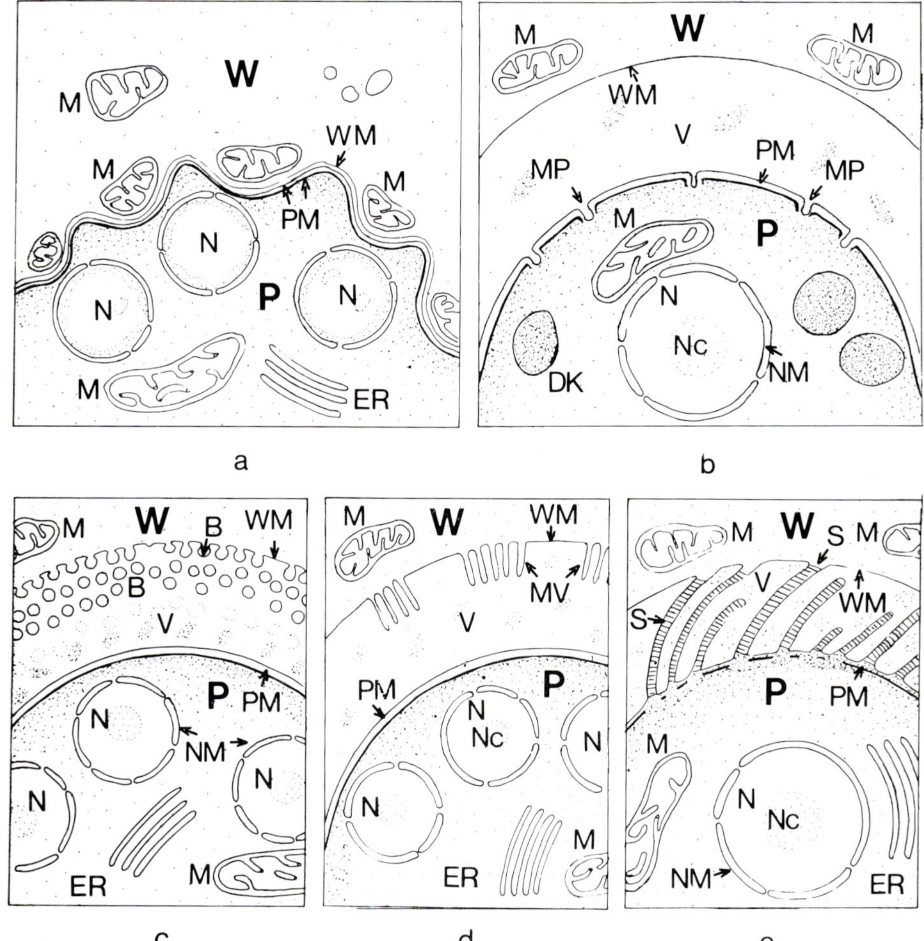

Figure 141. Electron-microscopic view of the **structure of the boundary between some Eimeria gamonts (Coccidia) and their host cell**

    a multinucleate microgamont of *Eimeria bovis* in the epithelium of the caecum and colon of cattle. Close contact between the membrane of the parasite (PM) and the inner membrane of the host cell (WM)

    b macrogamont (uninucleate) of *Eimeria stiedae* in the epithelium of the hepatic and bile ducts of the rabbit, with a parasitotrophic vacuole (V) forming a wide space between the parasite and the host membrane. Parasite membrane with micropores (MP)

    c microgamont of *Eimeria auburnensis* in the connective-tissue cells of the small intestine of cattle with migration of numerous small vesicles or vacuoles (B) from the host membrane into the parasitotrophic vacuole (V) organelle, for the penetration. The inner cell membrane assists in strengthening the parasitic cell, and the fibrils effect motility (figures 140, 1E, 4E, 7E, 10E). The cyclostome, which has already been laid down at this stage, is small and probably not yet functional. Shortly after penetration into a new host cell, the apical structures, the paired organelle, the inner cell membrane and the fibrils all disperse (figures 140 2E, 5E, 8E). The cytostome expands and food uptake begins. The structure of the cell nucleus changes, and nuclear division is accompanied by enlargement of the mitochondria, multiplication of ribosomes, and an increase in the endoplasmic reticulum. These changes are taken to be signs of increased protein synthesis.

Figure 140 8E represents a young trophozoite. In the course of nuclear division, the dispersed structures are again laid down. The first indications of a fresh production of these organelles are the thickening of the cell membrane as the inner membrane becomes established, and the appearance of the precursor of the paired organelle. Before they leave the cell, all merozoites are again equipped with the organelles necessary for penetration (figure 140 9E). The ookinete is surrounded by an outer and an inner cell membrane, and subpellicular fibrils are present. The

---

d microgamont of *Eimeria* sp. in the epithelium of the caecum and colon of the desert jerboa (*Dipodomys ordi*) with microvilli (MV) of the host membrane

e macrogamont of *Eimeria perforans* in the epithelium of the small intestine of the rabbit with tube-shaped canals (S), which link the parasite with the host cell across the parasitotrophic vacuole

B = vesicles (transport vacuoles) coming from the host membrane

DK = dark bodies

ER = endoplasmic reticulum

M = mitochondrion

MP = micropore of the parasite membrane

MV = microvilli of the host membrane.

N = nucleus

Nc = nucleolus

NM = nuclear membrane with pores

P = parasite

PM = double parasite membrane

S = tube-shaped connection of the parasite with the host cell (complete and partial longitudinal sections)

V = parasitotrophic vacuole

W = host cell

WM = host cell membrane surrounding the parasite

ookinete (figure 140 16, 16E) penetrates the stomach wall of the fly. Later it lies intracellularly. The oocyst is bound to the basal lamina of the epithelial layer and surrounded by a three-layered membrane. The cytoplasm includes mitochondria, abundant ribosomes and pigment inclusions in vacuoles, in addition to the cell nucleus. The oocyst cytoplasm, which is at first multinucleate, is divided up by the formation of splits and vacuoles. The sporozoites arise in these sporoblasts. The production of sporozoites starts with a thickening of the sporoblast membrane, and the nuclei lie in the immediate vicinity. The apical structures, the paired organelle, the inner cell membrane and the subpellicular fibrils are formed, and each sporozoite also contains mitochondria (figure 140 18E, 1E). Thus equipped, the sporozoite is able to invade the cells of the reticuloendothelial system after transmission. The formation of the organelles required for penetration proceeds in the same way in sporozoites, and in exoerythrocytic and intraerythrocytic merozoites. This process is always preceded by a thickening of the cell membrane at a narrowly defined site. Nothing is yet known about the mechanisms which trigger off the formation anew and retrogression of these structures, nor about the means by which these processes are co-ordinated.

Figure 140 shows that the different developmental stages of a protozoon, which have different functions, develop particular structures for these purposes at appropriate times. Comparative studies reveal, however, that there is by no means always only one structure possible for any given function. This is evident, for example, in the gamonts of the Coccidia in the genus *Eimeria* (compare figure 57l, n). These parasitic stages which live intracellularly take their food from the host cell and release products of their metabolism into it. The composition of the boundary surface between the parasite and its host cell is significant for this exchange of material. Figure 141 shows how varied the boundary structures for this function can be within a single genus, as they appear by electron microscope. Figure 141 a shows part of a microgamont of *Eimeria bovis*, in which the double cell membrane of the parasite lies closely applied to the inner membrane of the host. The parasite protrudes at places where the host cell has no mitochondria, thereby increasing its surface area. In figure 141 b the gamont of *Eimeria stiedae* lies in a large vacuole in the host cell which has formed as a result of the parasitic infection. Micropores in the membrane of the parasite evidently encourage exchange of substances. It is morphologically clear in the gamont of *Eimeria auburnensis* (figure 141 c) that substances from the host cell reach the parasitotrophic vacuole. The host cells form protrusions into the parasitotrophic vacuole, which are cut off as small vesicles and migrate into the vacuole while disintegrating. This process is reminiscent of the small transport vacuoles which carry substances into the cytoplasm from the food vacuoles (figure 129 h, 131 CNV). In a species of *Eimeria* of the desert jerboa, the host membrane is enlarged by numerous villous tubes, the **microvilli** (figure 141 d), while in *Eimeria perforans* (figure 141 e) tube-shaped structures form a connection from the parasite, through the parasitotrophic vacuole, direct to the host cell. These few examples give an insight into the multiplicity of the morphological solutions to a single functional problem in the realm of the ultrastructure.

Sexuality and heredity

The basic property of all living organisms is their capacity for reproduction while retaining their species-specific constitution. This capacity derives from the ability of the DNA molecules to increase by replication; since they contain the code for all the properties of the cell, these are the carriers of heredity, the genes. Since the morphology and functions of the cell are induced by the nucleus, nuclear division has particular significance in the propagation of the cell. Whereas sexuality and reproduction are closely linked and directly dependent on one another in the higher Metazoa, it appears in the Protozoa that replicative multiplication as the basic property of life by no means has to be instigated by sexual processes. Numerous Protozoa reproduce solely by asexual division, by agamogony, in which they themselves always live on in the developing daughter cells. In this sense, the protozoan cell is immortal.

When sexuality occurs, this is always characterized by karyogamy, the fusion of two haploid nuclei, although the form it takes may vary. These nuclei are called **gametic nuclei**. At syngony, the haploid gametic nucleus together with its own cytoplasm normally constitutes a gamete. The fusion of two

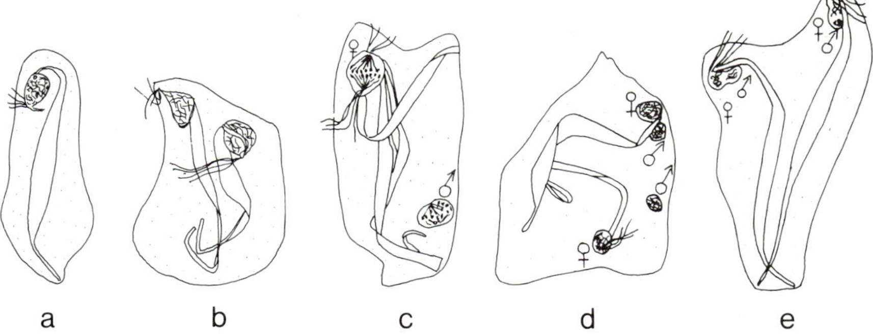

Figure 142. *Notila proteus* (Polymastigina). Fertilization without free gametes
  a Normal diploid flagellate with axostyle and 4 flagella
  b fusion of two diploid gamonts
  c meiosis of both nuclei, in which the male nucleus can now move freely
  d both male haploid gametic nuclei migrate to the female gametic nuclei
  e production of the zygotic nuclei through fusion of the gametic nuclei. Both axostyles are formed afresh after the previously united old axostyles have dispersed. Subsequently, two normal daughter animals form by cell division.

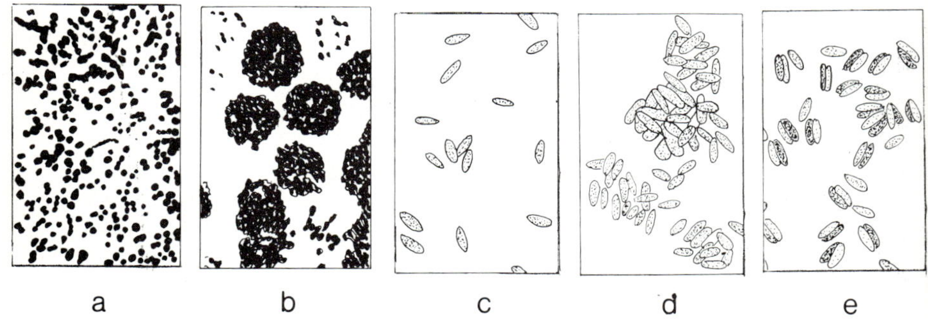

Figure 143. **Action of gamones**
a–b *Chlamydomonas eugametos* (flagellate of the Phytomonadina)
  a male gametes in uniform distribution
  b the same after addition of the filtrate of female gametes: isoagglutination
c–e *Paramecium bursaria* (holotrichous ciliate)
  c uniform distribution within one pairing type
  d clumping together 7 minutes after mixing with another pairing type
  e as d, but after 24 hours, with numerous conjugation pairs.

gametes produces the zygote. In just a few cases, actual gamete production does not occur. In *Notila proteus* of the Polymastigina (figure 142) and the amoeba *Sappinia diploidea* (figure 44), the gametic nuclei fuse within a common cytoplasm which was already there.

There are no free gametes in the conjugation of ciliates either (figures 98, 99); these only produce gametic nuclei, also called **prenuclei**. In this case, however, two gametic nuclei arise in each conjugant; one of each pair, the migratory nucleus, passes over into the other conjugant and forms a synkaryon with the stationary nucleus there. In all these cases the sexual process begins with two different cells, for which reason it is also called **amphimixis**. For this, it is necessary for two sexual partners to find one another. The affinity of the two partners for one another is determined by specific sexual hormones, the **gamones**. Figure 143 b shows the effect of these gamones in, for example, the flagellate *Chlamydomonas eugametos*; the male gametes agglutinate after the addition of a filtrate of female gametes. Carotinoids are known to be gamones. Since the ciliates are hermaphrodite, a female gynogamone cannot be differentiated from a male androgamone. On the other hand, conjugation types can be recognized in ciliates. Progeny of a single ciliate, i.e. ciliates of a clone, show no sexual reaction among themselves; such reaction is only possible with other suitable strains of the same species. If two suitable conjugation types are mixed together, there is a rapid agglutination from which the conjugation pairs subsequently emerge (figure 143 c–e). Whereas the gamones of flagellates can be secreted into the culture medium, the sexual hormones of the ciliate remain bound in the cell. They are induced in the cytoplasm by the macronucleus.

The sexual hormones can first occur and be functional in the gametes or even in the gametic mother cells. In the former case only the gametes find each other and a gametogamy takes place (figures 24, 25, 44, 53, 57, 58, 59, 60, 62, 67, 97). In gamontogamy, in which the gametic mother cells locate each other (syzygy), the reciprocal affinity of the gametes originating from them is, however, also retained from then on, as in gametogamy (figures 45 g, 46, 47, 48 n, 49 e, 51, 54, 55, 98, 99, 100). The behaviour of the Foraminifera (figures 44, 45) shows that gametogamy and gamontogamy can occur between closely related species. In the euciliates, only gamontogamy is possible,

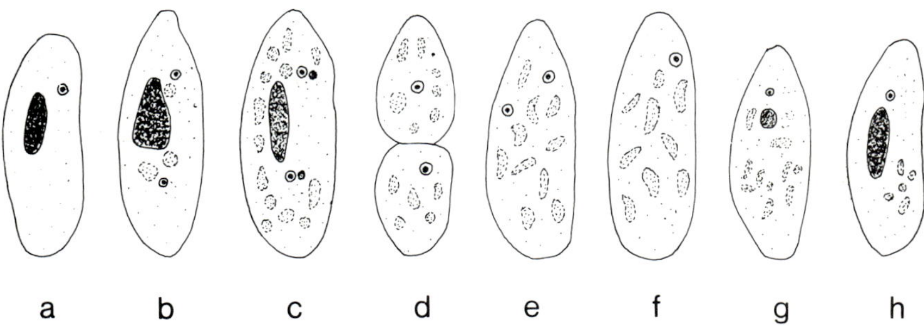

Figure 144. Diagram of **autogamy in a ciliate**
Four haploid micronuclei derived from the diploid micronucleus (a–c). Disintegration of the macronucleus, 2 micronuclei are shared between 2 daughter cells (d). After mitosis is complete, karyogamy occurs (e, f). The new micronucleus and macronucleus are formed through mitosis from the synkaryon (g, h).

in which the outward appearance of a gametogamy is simulated secondarily by total conjugation (figure 100); however, in true gametogamy each partner would only contain one gametic nucleus.

Syngamy in the form of oogamy corresponds to the sexual act in the higher Metazoa, but conjugation shows that a great multiplicity of sexual processes is possible in the Protozoa. This is also demonstrated by **autogamy**, which is carried out by a single protozoan cell without a partner and, as **automixis**, is contrasted with amphimixis. In the heliozoans *Actinosphaerium eichhorni* and *Actinophrys sol,* two daughter cells form in one outer coat and these fuse with one another as gametes (figure 44 f–i). In the foraminiferan *Rotalliella heterocaryotica*, the **gametes of a single individual fuse** (figure 45 q–s). Autogamy has also been demonstrated in some ciliates, e.g. *Paramecium bursaria, P. aurelia, Euplotes patella, E. minuta, Tetrahymena rostrata*; here, the nuclear processes take their course as in conjugation (figure 144). The haploid nuclei produced by meiosis are distributed between two daughter cells. Subsequent mitosis of these nuclei produces another 2 gametic nuclei in the daughter animals which fuse to form a diploid synkaryon from which the new nuclear apparatus develops, as in conjugation. Autogamy is called cytogamy when the individual is in purely superficial contact with another ciliate or with a conjugating pair.

In some cases, the further development of a gametic cell can proceed without fertilization. *Eucoccidium dinophili* provides an example of this **parthenogenesis**. At a low level of infection the macrogamete (figure 53 g) can undergo sporogony without fertilization. If there is no reciprocal fertilization of the daughter gametes in the heliozoan *Actinosphaerium* and *Actinophrys*, the viable gametes can continue to develop on their own. Occasionally in *Volvox aureus* of the Phytomonadina, too, an unfertilized macrogamete can lead parthenogenetically to the formation of a colony.

Since all processes of life depend for their specificity on the species-specific DNA of the cell, its distribution in reproduction and sexuality has no great significance. Thus the division of a cell is always limited to nuclear division. In this, the DNA of the nucleus collects in the chromosomes which, together with proteins, are structures which are easily demonstrated by staining. Figure 145 shows these processes in the gregarine *Monocystis rostrata* (see figure 47 a) where 8 chromosomes are formed (figure 145 a, b). To ensure that both daughter nuclei contain the original amount of DNA, there is a doubling of the chromosomes when forming the chromatids, which is linked with unwinding and new formation of the double helix of the DNA molecule (c). This produces 2×8 chromosomes which are divided between the daughter nuclei (e–g). If a haploid gametic nucleus forms from the normal diploid nucleus after syzygy of two gregarines, the 8 chromosomes are divided up as 2×4 chromosomes in meiosis (h–l). The processes of nuclear division have a corresponding pattern in other diploid nuclei too, e.g. in the micronuclei of the ciliates *Kidderia mytili* and *Didinium nasutum* (figure 146 a–e). The reduction of the chromosome number in meiosis is necessary so that the zygote can contain the normal diploid chromosome complement. The question of the point at which meiosis takes place in this alternation between the diploid and the haploid condition has already been referred to in figure 103. It determines whether the normal vegetative phase of a protozoon is diploid or only haploid.

The separation of the chromosome complement in mitosis and meiosis generally takes place in the field of force of a division

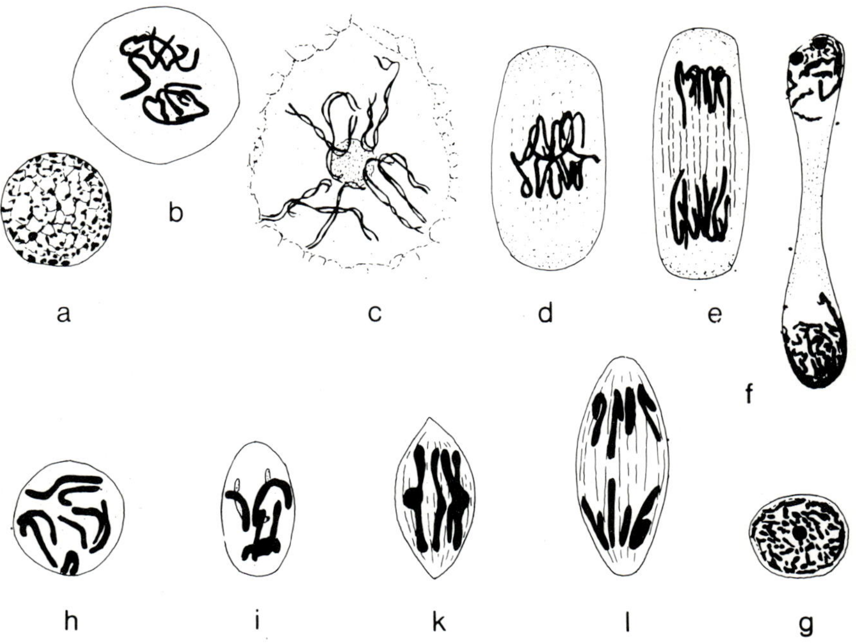

Figure 145. **Nuclear divisions** in the gregarine *Monocystis rostrata*
a–g mitosis
h–l meiosis
    a interphase nucleus
    b prophase with 8 chromosomes in the diploid nucleus
    c prometaphase with paired daughter chromosomes, the chromatids, and dissolved nuclear coat
    d metaphase with arrangement of the chromosome pairs in the equatorial plane
    e anaphase with separation of the two diploid sets of chromosomes
    f–g telophase with division of the nucleus and production of two daughter nuclei with loosening of the chromosomal structure
    h diploid nucleus of a gamont in syzygy with 8 chromosomes
    i paired arrangement of the 8 chromosomes
    k onset of separation of the pairs of chromosomes with 4 chromosomes going each way
    l haploid sets of chromosomes drawing apart from one another to give 2 haploid gametic nuclei after nuclear division (as in f).

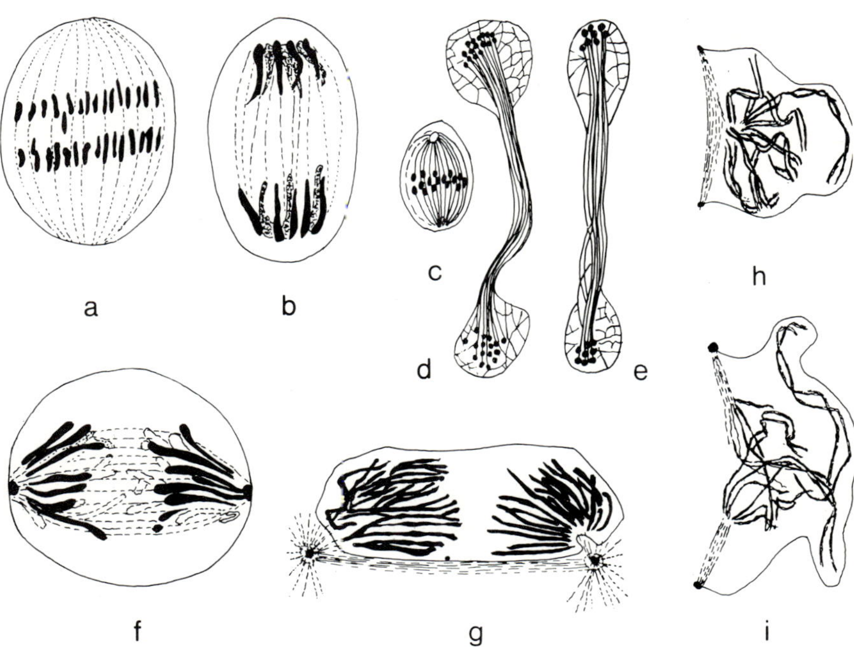

Figure 146. **Distribution of the chromosomes** during nuclear division
a–b *Kidderia mytili* (thigmotrichous ciliate)
  a mitosis of the diploid micronucleus with 16 chromosomes going each way (early anaphase)
  b meiosis with two sets of 8 chromosomes (late anaphase)
c–e *Didinium nasutum* (holotrichous ciliate)
  c micronucleus with diploid set of chromosomes (16 chromosomes)
  d mitosis with two sets of 16 chromosomes
  e meiosis of the micronucleus with two sets of 8 chromosomes
  f nucleus of **Pyrsonympha** (polymastigine flagellate) in anaphase with an intranuclear division spindle
  g nucleus of **Pseudotrichonympha** (polymastigine flagellate), anaphase with extranuclear spindle formation
  h–i *Aggregata eberthi* (coccidian, *Eimeridea*), mitosis without production of an equatorial plate in metaphase = paramitotic type of division

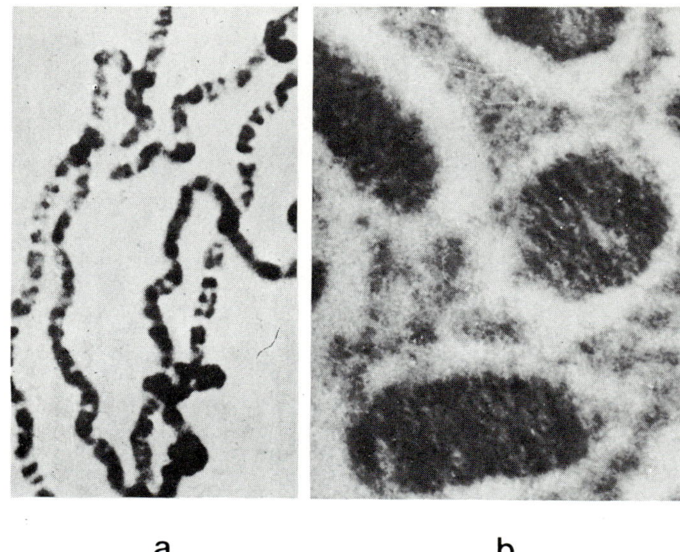

a   b

Figure 147. **Chromosomal structures**
a *Holomastigotoides psammotermitidis* (polymastigine flagellate), chromosomal structure seen under the light microscope during early prophase (stain: carmine acetic acid) (magnification 2290 ×)
b *Amphidinium elegans* (dinoflagellate), section of chromosomes in electron microscope with internal tight-spiral structure (magnification 71,440 ×)

spindle between two centrioles. They arise from the division of an existing granular centriole usually lying in the cytoplasm at the periphery of the nucleus, and eventually come to lie at opposite poles. In many cases the development of a special area around the centriole leads to the formation of a centrosome (figure 146 g). Centrioles and centrosomes function as a **kineto centrum** during the separation of the chromosomes. In many Protozoa the centrioles enter the nuclear region and form an intranuclear division spindle (figure 146 f); in this case, the nuclear membrane breaks up sooner or later during division. In the Tetramitidae and Hypermastigidae, an extranuclear division spindle develops (figure 146 g) which participates in nuclear division only after this has begun. Since in this case the nuclear membrane is retained for a long time during mitotic chromosome division, the chromatids draw away from one another under the influence of special pulling-threads which emanate from a

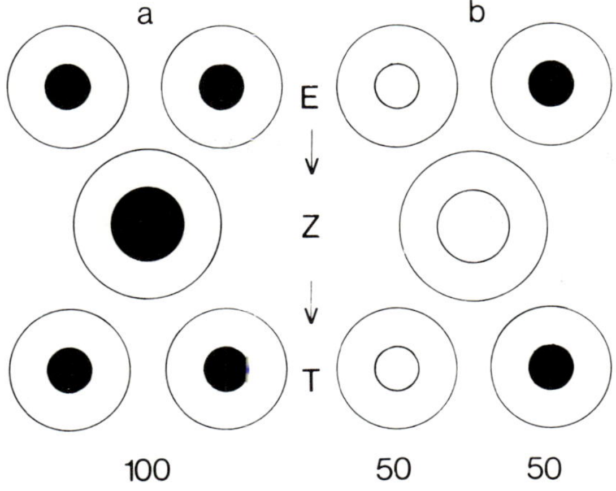

Figure 148. **Inheritance of a chromosomal pair in haplonts**
  E = parents
  T = daughter cells (gones)
  Z = zygote
  a union of similar parents (black) produces the homozygous black zygote
  and 100% similar black gones
  b crossing of dissimilar parents (white and black) produces the heterozygous white-black zygote, and the gones derived from it are 50% white
  and 50% black

centromere lying by the nuclear membrane. The chromosomes
generally become arranged in the equatorial plane at the
beginning of nuclear division (figures 145 d, 146 a, c) but this
stage can be omitted in some cases (figure 146 h, i). This
peculiarity is also called **paramitosis**.

The spindle attachment, the **kinetochore**, can sometimes be
seen under the light microscope in stained preparations with
large chromosomes (figure 151 a–d). A spiral structure is clear in
the large chromosomes of some Polymastigina (figure 147 a).
With the aid of phase contrast microscopy, the mitotic processes
can be seen in some of these flagellates, even while they are still
alive. Electron-microscopic study of protozoan chromosomes
has not contributed anything very significant; in section, only
an internal tight-spiral structure is visible (figure 147 b).

After two haploid gametic nuclei have united to produce a
synkaryon, the zygotic nucleus contains maternal and paternal

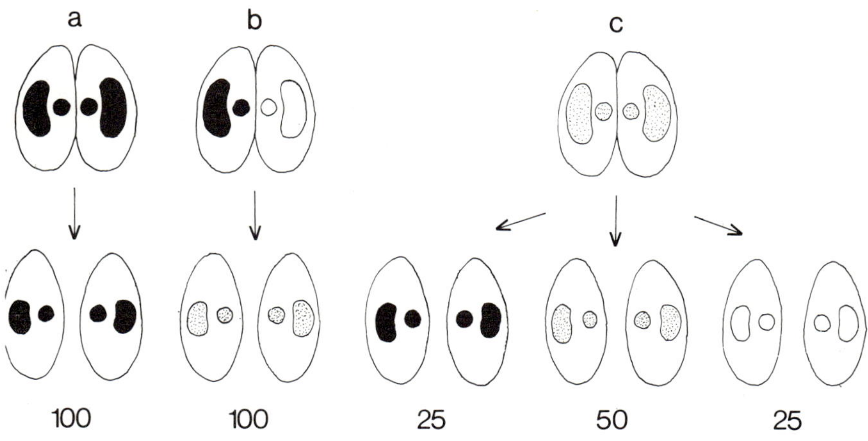

Figure 149. **Inheritance of a chromosomal pair in the conjugation of diploid ciliates**
    a conjugation of two similar homozygous ciliates produces 100% similar daughter animals
    b crossing of two homozygous dissimilar ciliates (black and white) produces 100% heterozygous daughter animals
    c conjugation of two heterozygous ciliates produces 25% homozygous black, 25% homozygous white and 50% heterozygous black-white

chromosomes so that characteristics of both parental cells are transmitted to the zygote. This **inheritance** in Protozoa is subject to the Mendelian laws operating in higher plants. If the genes of the chromosome of a haploid gametic nucleus match those of the corresponding chromosome of the other gametic nucleus, so that two similar alleles form a homologous chromosomal pair during synkaryon formation, the resultant diploid nucleus is homozygous with respect to this chromosomal pair (figure 148 a). Meiosis in this zygotic nucleus then leads to 100% similar gametic nuclei or gones. If, however, the chromosomal structures corresponding to each other embody different characteristics, the zygotic nucleus contains both attributes; it is heterozygous (figure 148 b). Meiosis then leads to a 50% distribution of the characteristics of each parental cell in the gametic nuclei. The path of inheritance is more complex in the diploid Protozoa, although the same rule applies. If both alleles of a chromosome pair are similar to each other and to those of the sexual partner, i.e. if, for example, the two conjugants

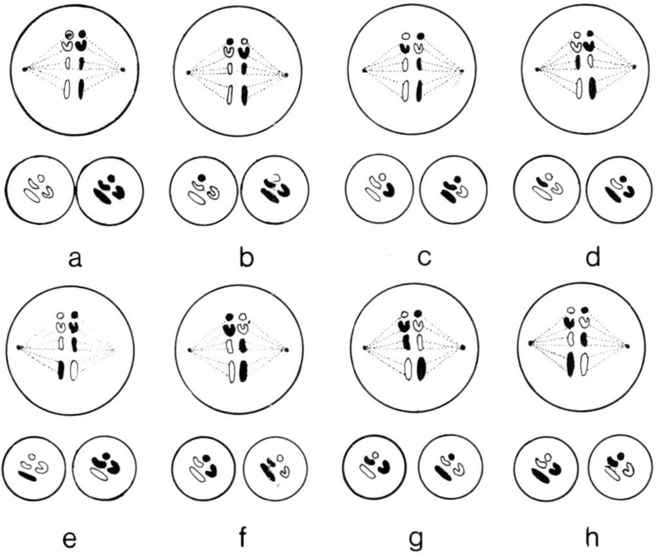

Figure 150. The 4^2 **possible distributions of 4 maternal and 4 paternal chromosomes** in meiosis = 16 different gones.

in the ciliates are similarly homozygous, conjugation produces 100% similar homozygous offspring (figure 149 a). If both conjugants are still homozygous but dissimilarly homozygous as in figure 149 b (2 black alleles on one hand and 2 white alleles on the other), crossing, when the next generation of ciliates each inherit one allele from each conjugant, leads to 100% heterozygous ciliates. In this case a mixture of the characteristics can appear, unless one of the inherited characteristics is a recessive one with its effect masked by the other (dominant) one. If two heterozygous ciliates are crossed with one another, according to Mendelian rules 50% of the daughter cells will be homozygous, namely 25% homozygous black and 25% homozygous white (figure 149 c), while the remaining 50% will be heterozygous. Since, however, one genome in the Protozoa always comprises more than one chromosome (compare page 188), inheritance of characteristics is much more complicated than this. Figure 150 shows this for a zygote with 4 pairs of chromosomes in which the alleles are different (white

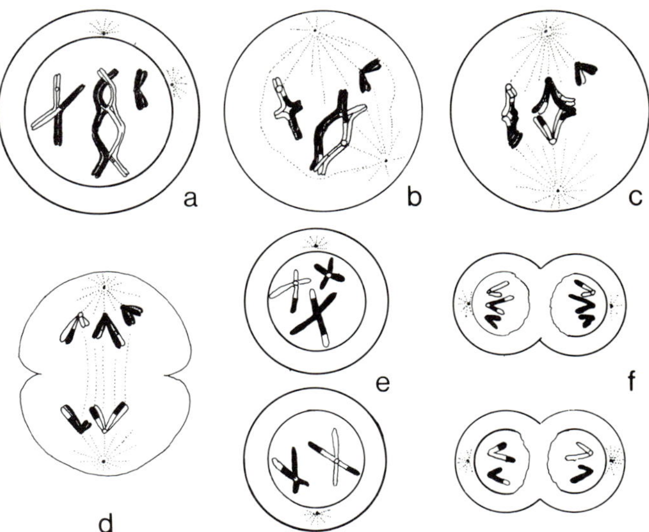

Figure 151. Diagram of **exchange of parts of chromatids** in a 2-step meiosis

a Prophase with paternal and maternal chromatids in the diploid nucleus
b prometaphase with exchange of pieces of chromatid between the paternal and maternal chromatids
c metaphase
d anaphase
e daughter cells after the first step
f 4 different haploid gones after meiotic division has taken place (second step) in the daughter cells (e); 3 of the gones have a chromosomal structure which has undergone recombination

and black). There are 16 possible combinations for the haploid nuclei produced by meiotic division. The number of these possibilities can, however, be raised even further if an exchange of pieces occurs between chromatids during meiosis (figure 151). This produces chromosomes which no longer simply correspond to the paternal or maternal type, but which have exchanged individual genes with one another, so that the resultant chromosomes now represent a partial mixture.

Changes in characteristics can also occur in autogamy if the nucleus is heterozygous. The principle of autogamy in the ciliates can be seen from figure 144. It gives a distribution of 50% homozygous black and 50% homozygous white when the

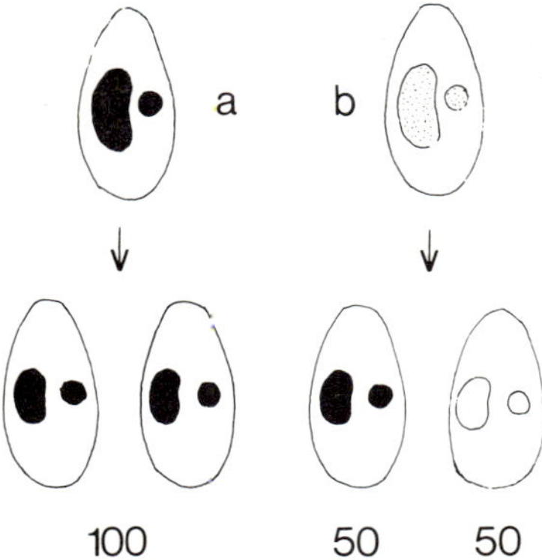

Figure 152. **Inheritance of a chromosomal pair in the autogamy of a ciliate**
　a autogamy of a homozygous ciliate produces 100% similar homozygous daughter cells
　b autogamy of a heterozygous ciliate produces 50% homozygous black and 50% homozygous white

ciliate is a heterozygote (black-white, figure 152b). The progeny of a homozygous ciliate (figure 152a) are homozygous and similar to the parent.

The prerequisite for all new combinations is the occurrence of differences between the genes of the sexual partners. Such differences arise occasionally through spontaneous changes in the genes, through **mutations**. These changes in the genes can also be induced experimentally in Protozoa by the influence of mutagenic substances, such as urethane and antibiotics, but particularly by irradiation with X-rays or ultra-violet light. These changes are immediately effective in haplonts in which, of course, not all mutations are viable, and the few viable mutations must first be grown on further. In diplonts, a mutation in one allele can be suppressed or moderated if it is recessive to the other allele. The recognition of a mutation in the generative micronucleus of ciliates is impeded by the fact that the somatic character is determined by the polyploid macronucleus. In this case, only autogamy will reveal the mutation since here, according to figure 152, homozygous offspring, each with a new macronucleus, are produced.

The entire gene complement produces the genetically deter-

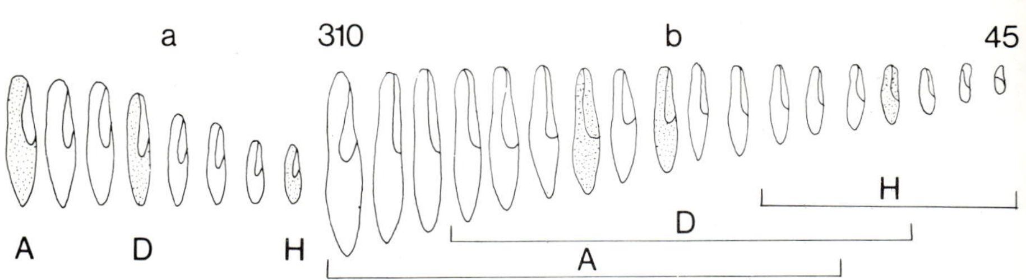

Figure 153. *Paramecium caudatum*
   a average lengths of **8 races** (A–H) from one population. Races A, D and H stippled
   b Variations in size in the clones of the races A, D and H. The average in each case is stippled. Sizes of the two extremes: 310 μm and 45 μm.

mined developmental potentialities of a protozoon, the **genotype**. The sum of characters based in the hereditary material can, however, develop differently within genetically determined limits, under the influence of conducive or inhibiting environmental factors. The **phenotype**, the actual manifestation of character, is a **modification** occurring at that particular time. This is easiest to see by considering size. Figure 153 a shows eight races of *Paramecium caudatum*, which have been produced from the population of a single expanse of water by the establishment of clones. The figure reproduces the average sizes. Figure 153 b shows the whole spectrum of modifications for the stippled races A, D and H, which show a normal (Gaussian) distribution, the average size (stippled) being the most frequent in the clone. The genetically determined range of modifications of the race shown here is retained, irrespective of the size of the individual within a pure race from which the new clones are produced. The modifications controlled by environment also include the respective phenotypes of many parasites. *Leishmania* species of the Trypanosomatidae produce the aflagellate amastigote (leishmania-form) (figure 18 d) in the tissues of their vertebrate host or in tissue cultures at 37 °C. The promastigote flagellate form (figure 17 b) arises from the amastigote stage in the insect vector, or in culture media containing blood at room temperature. In the same way, many trypanosomes alternate between the trypomastigote stage (figure 18 c) in the blood of their vertebrate host and the epimastigote form in the gut of the vector (figure 17 c). Here, each phase can be retained under experimental conditions for as long as is required; this shows that there is no obligatory rhythm, but really only a dependence on the prevailing environmental conditions. The amoeboid form of *Naegleria gruberi* can change very quickly into the flagellate stage when food is scarce (figure 17 k, l).

The Protozoa show that life can begin and continue even without sexuality, and that sexuality represents a process acquired only secondarily in the evolution of living organisms; it assumes an ever-increasing significance as they become more highly developed. However, even Protozoology reveals the significance which sexuality can assume. It facilitates a much greater potentiality for variation in all characteristics within a species; by this means, although many misdevelopments arise, some genotypes are produced which, according to **Darwin**, are better equipped for the struggle for survival.

The examples of inheritance shown so far are all dependent upon the DNA of the nuclear chromosomes, upon the **genome**; i.e. they are karyotically determined. However, some cases of inheritance in the Protozoa are dependent upon extrakaryotic DNA. This extrakaryotic DNA occurs, as already mentioned, in the plastids of the phytoflagellates, in the mitochondria (including the kinetoplasts) and also in the centrioles. In contrast to the genome, all the genes of the plastids comprise the **plastome** and the genes of the mitochondria the **chondriome**. The proportion of this DNA is usually small in relation to the nuclear DNA. In the flagellate **Euglena** the plastid DNA constitutes 1%, and in **Chlamydomonas** gametes over 6%. DNA can also be present in the cytoplasm of Protozoa because of symbiotic parasites.

In the discussion of plastids and mitochondria it has already been established that their DNA is not quantitatively sufficient for an autogenous development of these organelles, so that these must be coded in addition by the nuclear DNA. This applies also to the killer property of paramecia which, as electron microscopy shows, can be attributed to bacterial cytoplasmic inclusions, the **kappa symbionts**. These symbionts, easily seen with Giemsa staining, are dependent upon the presence of a gene K in the nucleus, whereas the allele k does not permit the development of the symbionts. Paramecia which contain the

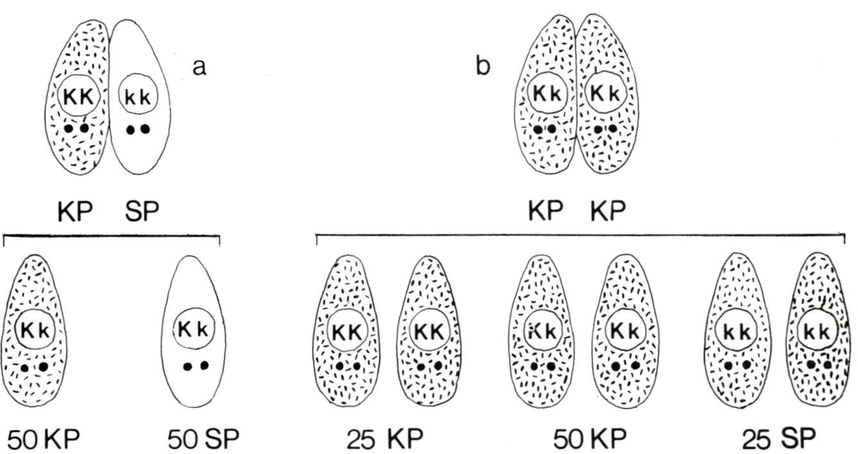

Figure 154. **Inheritance of the** cytoplasmic **killer-character**, dependent on the gene K

KP = killer paramecium

SP = sensitive paramecium

a The conjugation of a homozygous killer paramecium (KK) with a homozygous sensitive paramecium (kk) produces 50% heterozygous killer paramecia (Kk) and 50% heterozygous sensitive paramecia (Kk)

b The conjugation of two heterozygous killer paramecia produces 25% homozygous and 50% heterozygous killer paramecia and 25% homozygous sensitive paramecia

kappa symbiont in their cytoplasm are called killer paramecia (KP) because they kill other paramecia which lack the kappa symbionts by means of toxic secretions. The kappa-free paramecia are hence called **sensitive paramecia** (SP). Figure 154 shows the inheritance of the cytoplasmic kappa infection. Sensitive paramecia can cross with killer paramecia (figure 154 a) because they are resistant to the killer toxin during conjugation. The genes K and k are subject to the Mendelian rules, whereby the conjugation of a homozygous K and a homozygous k leads to 100% heterozygous k(Kk (compare figure 149 b). Nevertheless, crossing produces only 50% killer paramecia, and the other 50% are sensitive since during conjugation it is usually only the migratory nucleus, without cytoplasm and without kappa symbionts, which passes over into the sensitive partner. If two heterozygous killer paramecia conjugate (figure 154 b), according to Mendelian inheritance (figure 149 c) they produce

50% heterozygous and 25% homozygous killer paramecia. Although the remaining 25% homozygous paramecia with the genes kk at first contain kappa symbionts, they are sensitive paramecia since the kappa symbionts perish without the gene K.

The amoeba *Pelomyxa palustris* (figure 27 a) provides an example of the great significance the DNA of a symbiont can have for the protozoan cell, in contrast to the previous example. Since this amoeba has no mitochondria of its own, the symbiotic bacteria here take over the function of mitochondria. The possible results of crossing, cited with respect to the killer character, lead to the important finding that the extrakaryotic inheritance is not subject to Mendelian laws. This applies basically to all extrakaryotic inheritance, and hence also to the characters coded by the plastid DNA of the phytoflagellates. The plastid DNA, like the nuclear DNA, carries the information for protein synthesis; its own ribosomes, unlike the normal ribosomes of the cytoplasm, are involved in this synthesis. Antibiotics cause deleterious changes in ribosomal structure. In a mutant of *Chlamydomonas rheinhardi* which has been subjected to 500 µg streptomycin/ml, the inheritance of increased resistance reveals that an extrakaryotic inheritance dependent upon the plastid DNA is functioning here. Another mutant of *Chlamydomonas*, which is characterized by an increased resistance to heat, and in which there must therefore be a modified protein structure, is coded by the plastid DNA. But an extrakaryotic mutation can be dependent not only on the DNA of the chloroplasts, but also on the genes of the mitochondrial DNA. There is a mutation in *Chlamydomonas rheinhardi* which is not only resistant to streptomycin but which is actually streptomycin-dependent; this can be traced to the presence of sd genes in the mitochondria. If these streptomycin-dependent strains are transferred to a medium lacking streptomycin, mutation in the mitochondria can lead to a reappearance of the streptomycin-sensitive ss gene. Selection leads to the formation of streptomycin-sensitive colonies in this medium. They are, however, at first only phenotypically sensitive to streptomycin, since to begin with there are always a few remaining sd genes. All the sd genes are finally lost only after a longer time in the streptomycin-free medium. Only then is a gradual transference back to a medium containing streptomycin no longer possible. As long as some sd genes are retained, the strain can re-adapt to a medium containing streptomycin.

This behaviour corresponds to the phenomenon of "**Dauer-modifikation**" with the protracted retention of an acquired characteristic. These long-lasting modifications are also known in the pathogenic trypanosomes, which can become resistant to drugs during therapy with arsenical preparations; experiments with animals show that this character can be retained, even without the influence of further medication, for months or years before it is reversed. Long-lasting modifications like this also arise against antibodies. In *Paramecium caudatum*, adaptations made towards arsenicals show that here the increased resistance is evidently linked to the alleles of the somatic macronucleus. When a new macronucleus is formed from the micronucleus after a conjugation, the resistance to arsenic is immediately lost. On the other hand, resistance to calcium in paramecia is not influenced by conjugation or autogamy.

Cell associations and cellular differentiation

It was said at the outset that the word "protozoon" does not necessarily imply a unicellular condition. If the Protozoa are nevertheless generally unicellular, this is based on the nature of their organization. They do not form the physiological unity of a centrally controlled multicellular confederation. However, there are the first indications of some tendencies towards a cell association. Colonial associations occur in the flagellates Chrysomonadina (figures 2 g, h, l, m, 3 c, f), Phytomonadina (figures 14, 15) and Protomonadina (figure 16 b, c, d). Some genera of the Radiolaria (**Collozoum, Sphaerozoum, Collosphaera**) form 4–6 cm swimming colonies. Many peritrichous ciliates, too, form colonies (figure 88). These associations generally provide greater protection to the individual cells and increase the water current and the procuring of food by means of the action of the flagella and cilia. The protective effect is increased still more when, as in **Carchesium** and **Zoothamnium**, a contact stimulus can be transmitted by one individual to the entire colony, thereby leading to a united escape reaction by contraction. A further advance conducing to organism survival is cell differentiation. This occurs temporarily in some amoebae, the **Acrasiae**, for example in *Dictyostelium discoideum* (figure

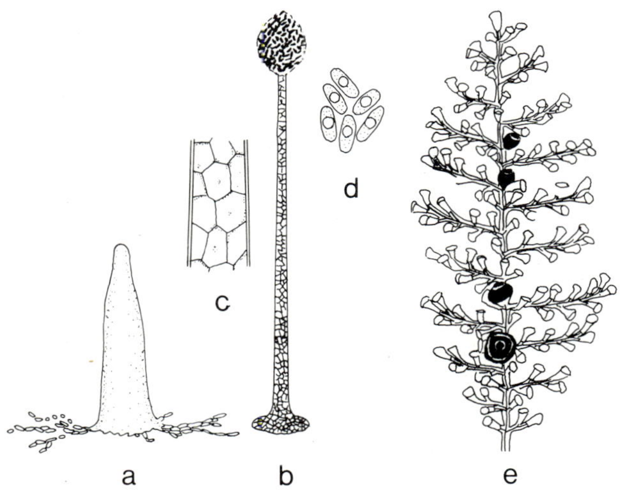

Figure 155. **Formation of colonies**
a–d *Dictyostelium discoideum* (amoeba figure 28 g)
  a Formation of conus after aggregation
  b sporophore (= cyst bearer) with stalk (**c**) and cysts (**d**)
  c and **d** more highly magnified
  e *Zoothamnium alternans* (peritrichous ciliate). Colony with microzooids
(clear) and macrozooids (dark)

28 g). When environmental conditions are unfavourable, individual amoebae secrete a hormone, acrasin, which diffuses into
the medium and stimulates the amoebae to mass together into a
conus structure (figure 155 a). The conus tips over; all the
constituent amoebae shift their position, and finally pile up
again to form the so-called **sporophore** (figure 155 b). Cell
differentiation occurs in this structure; the amoebae constituting
the stalk fuse together into a compact tissue (c), while the
amoebae in the head of the sporophore survive as cysts (d).

In the flagellates, the individuals forming an association of
cells usually retain their capacity for independent development.
However, in the Phytomonadina, some Volvocidae show differentiation. In figure 14 the individual colonies contain cells which
are still all the same and which can all divide, but in
Pleodorina californica (figure 15 b) there are two types of cell
which can be distinguished by their difference in size. The
small somatic cells have lost the capacity to divide, so that
only the larger generative cells are still capable of forming new

colonies. In *Volvox aureus* and *V. globator* (figure 15 c, d) only a few generative individuals remain. A corresponding condition is found in the peritrichous ciliate *Zoothamnium alternans* (figure 155 e). Only a few macrozooid individuals can detach themselves and are still able to divide, while the many somatic microzooids have lost the ability to produce daughter colonies.

The **Cnidosporidia** have a particular form of cell differentiation during the production of spores. The site formation of the spores in Myxosporidia is multinuclear at first, and becomes temporarily multicellular (figure 66 a, b). In this relatively short phase of development, the physiological unity related to that of the metazoan cells is still less marked than in, for example, the colonies of the Phytomonadina.

In the Phytomonadina (Volvocidae) not only agamous reproduction but also sexuality is sometimes limited to a few individuals scattered within the colony. This differentiation into pure mortal soma cells and the cells of the germ path achieves its highest development in the genus **Volvox**. A parallel intracellular development occurs in the macronuclei of the euciliates and of some Foraminifera which also have only a somatic function. These examples reveal a certain tendency in the Protozoa towards the subsequent development of the Metazoa, but the fact remains that they do not yet have the physiological unity of a multicellular organism as possessed by the true Metazoa.

THE PROTOZOA AND THEIR ENVIRONMENT

The Protozoa and their Environment

The habitat of the free-living Protozoa

Protozoa can develop only in a habitat with sufficient moisture. Free-living **marine Protozoa** include flagellates, rhizopods and ciliates. These include the flagellate Coccolithophoridae (Chrysomonadina, figure 3 g, h) which are a chief constituent of marine plankton, as well as being the smallest planktonic Protozoa in the sea. Their calcified coats, the coccoliths, constitute a substantial proportion of marine sediment and of chalk. On the other hand, the Silicoflagellata (figure 2 o, p), likewise assigned to be Chrysomonadina, have siliceous skeletons, whose deposits can be seen in chalk and in Silurian silica slate. The dinoflagellates, too (figures 5, 7, 9), are widespread in marine plankton in an enormous profusion of different forms and are very important as a primitive food for many animals. While some species are restricted to certain climatic zones, others (such as *Noctiluca miliaris*) have a world-wide distribution. The Radiolaria (figure 34) and Acantharia (figure 35), which belong to the Rhizopoda, similarly constitute an important component of marine plankton. The majority of the Radiolaria live pelagically to depths of 350 m. Some radiolarians, however, thrive at great depths in the sea, as much as 4000 or 5000 m deep according to species. The Radiolaria, too, include warm-water and cold-water forms. The sedimented silica skeletons of the individuals which have died, the radiolarites, can in some places cover large areas of the sea bed in a thick layer. Where volcanic activity has caused a large amount of silicic acid to reach the sea, as in the region of the Barbados islands, the radiolarians can develop particularly abundantly. Fossil radiolarians have been found representing the Precambrian period onwards. Of the Foraminifera (figures 39–42) which belong to the Rhizopoda, only a few species such as the *Globigerina* species (figure 41 f) are planktonic, mainly in warm seas to a depth of about 100 m. By far the larger proportion of them belong to the benthos. These Foraminifera are attached to a substratum, or move along only slowly by means of their pseudopodia. The Foraminifera, too, can be divided into cold-loving and warmth-requiring species. The limestone cases of the forms occurring close to the surface of the sea are particularly durable. The fossilized cases of the Foraminifera are very significant in the dating of layers in palaeontological borings. *Fusulina* in the Carboniferous period and *Nummulites* are particularly important for this. The Foraminifera are a component part of solid rocks, but also of Jurassic, Cretaceous and Tertiary clays and marls.

Among the ciliates, the highest incidence of marine forms occurs in the oligotrichous *Tintinnidium* spp. (figure 84), only a few of whose species live in fresh water. They thrive in ocean plankton or mesoplankton in the upper layers of the sea. The oldest *Tintinnidium* remains found date from the Jurassic. Fossil *Tintinnidium* limestones occur together with coccoliths, particularly on the island of Mallorca and the rock strata of North Africa across the Alps as far as Iraq, but also in California and Australia. Finally, among recent marine ciliates worth a mention, the homokaryotic genus *Stephanopogon* (figure 73 e) and the holotrichous euciliates with a diploid macronucleus (figure 74 f) occur in the mesopsammon, the interstitial system of sea sand. Some marine Protozoa can become acclimatized to high concentrations of salt, up to 10% and higher, by experimental exposure. A natural adaptation of this sort occurs in Protozoa from salt pits, e.g. in the flagellates *Bodo caudatus*, *Asteromonas gracilis*, etc., in *Amoeba salina*, and in the ciliates *Frontonia marina* and *Fabrea salina*.

Fresh-water Protozoa live under extremely varied environmental conditions which can range from clear springs to putrescent sewage. Temperature, salinity, incidence of light, current and pH-value are important factors which determine the respective habitats for the individual species. However, the level of purity of the water has an even greater influence on the protozoan fauna; this is an index of the degree of contamination with organic matter which leads to bacterial decay and which influences the oxygen content of the water. The **polysaprobes** live in waters with a high level of protein decomposition. These include, from the Protozoa mentioned earlier in the taxonomic section, *Euglena viridis* (figure 8 c), *Polytoma uvella* (12 c), *Bodo putrinus* (see 17 e), *Trimastigamoeba philippinensis* (16 e, f), various limax amoebae (see 28 a, 29 a), also *Pelomyxa palustris* (27 a), and the ciliates *Colpidium colpoda* (77 c), *Paramecium putrium*, *Caenomorpha medusula* (82 k) and the suctorian *Sphaerophrya soliformis* (see figure 91 d). The **mesosaprobes** grow in waters in which oxidation processes occur as well as the breakdown of organic matter. They constitute a rich protozoan fauna. They include among the flagellates, most *Euglena* species, *Astasia* and *Peranema trichophorum* (figures 8, 10), many species of the genus *Cryptomonas* (6 a), *Chilomonas paramecium* (6 b), various *Chlamydomonas* species (13 d), *Gonium pectorale* (see 14 c), *Synura uvella* (2 g), *Bodo saltans* (17 e), the rhizopods *Chaos diffluens* (*Amoeba proteus*) (figure 26 a), *Astramoeba radiosa* (27 d), *Thecamoeba verrucosa* (27 e), *Euglypha alveolata* (37 h), *Actinosphaerium eichhorni* (32 a), the ciliates *Paramecium caudatum* (77 b), *Paramecium bursaria* (77 d), *Colpoda cucullus* (76 e), *Uronema marinum* (78 b), *Chilodonella cucullus* (74 h), *Lionotus fasciola* (75 g), *Urotricha farcta* (75 e), *Urocentrum turbo* (77 e), *Spirostomum ambiguum* (82 a), *Stentor coeruleus* (101 e) and *S. polymorphus* (82 d), *Didinium nasutum* (74 b), *Coleps hirtus* (75 d), *Oxytricha fallax* (86 a), *Aspidisca lynceus* (86 g), *Vorticella campanula* (88 a) and *Carchesium polypinum* (88 e). The **oligosaprobes** thrive in water which has a poor supply of organic matter and is rich in mineral substances. They include, for example, *Ceratium hirundinella* (5 d), species of the genus *Dinobryon* (21), *Eudorina* (14 e), *Volvox* (15 c, d), *Chromulina* (2 b) and *Mallomonas* (3 b), heliozoans of the genus *Acanthocystis* (33 g), the ciliates *Nassula elegans* (74 g), *Dileptus anser* (74 e). All the species listed can act as biological indicators of the respective degree of contamination of the water.

Mosses constitute a suitable habitat for some Testacea (figures 36–38), e.g. for the genera *Difflugia*, *Euglypha* and *Centropyxis*, but also for *Nebela* and for some flagellates and small ciliates. Before Protozoa can be called **Soil Protozoa**, their vegetative form must be found in the soil, since the cysts of all free-living Protozoa can be seen in soil samples. Since the soil Protozoa are often only sparsely distributed, they are concentrated by culturing. The presence of the vegetative stages can be ascertained, although not infallibly, from the difference between the results of culture of soil samples which have been treated with 2% hydrochloric acid, so that only the protozoan cysts still survive, and the results of culture of untreated soil samples. The following genera have been demonstrated by this means: the flagellates *Euglena*, *Astasia*, *Peranema*, *Oicomonas*, *Monas* and *Bodo*, and the rhizopods *Euglypha*, *Arcella*, *Difflugia*, *Centropyxis*, *Pelomyxa palustris*, *Thecamoeba verrucosa* and limax amoebae, and numerous ciliates such as *Paramecium*, *Colpoda*, *Nassula*, *Urotricha*, *Lacrymaria*, *Chilodonella*, *Colpidium*, *Glaucoma*, *Pleuronema*, *Uroleptus*, *Oxytricha*, *Urostyla*, *Stylonychia*, *Aspidisca* and *Vorticella*.

Examples of strange habitats for Protozoa are frog-hopper froth (cuckoo spit) for ciliates, and the surface of glaciers of the polar regions and high mountain chains for the flagellate *Chlamydomonas nivalis*, which causes the phenomenon of "bloody snow" by the production of haematochrome.

The Protozoa as parasites and causal agents of diseases

In contrast to the free-living Protozoa, the parasitic species live permanently or for a time in or on another living organism, the host. Only a few Protozoa have a plant host; nearly all the hosts of parasitic Protozoa are animal organisms which range from other Protozoa to the most highly developed mammals including man. If the host organism suffers no recognizable injury as a result of infection by the parasite, the parasite is called **commensal**. If the host is injured or even killed, we refer to a **pathogenic parasite**, the causal agent of a disease. If the parasite and host receive mutual benefit from the association, this is a **symbiosis**. Since most parasites also perish if their host dies, a high degree of pathogenicity can be regarded as a very incomplete adaptation between parasite and host. Hence a potentially pathogenic parasite must not always be pathogenically effective. The pathogenesis of a parasitic infection depends on the virulence, the respective degree of pathogenecity, of the parasite and the resistance of the host. The resistance of the host may be an innate species–specific natural resistance or may be acquired only under the influence of infection. In this latter case it is called **immunity**. The defence of the host can be directed either against the toxic effect of a parasite, or against the parasite itself, and then takes the form of either resistance to toxins or of reduction or elimination of the parasitic infection. This is achieved either by the production of antibodies which are directed specifically against the antigens of the parasite and of its secretions, or by cellular defence reactions which usually lead to the killing or phagocytosis of the parasite. The demonstration of specific antibodies in the blood of the host is very useful in the diagnosis of the infection.

The reactions of the host represent a permanent change in the environment of the parasite, to which the parasite for its part must react in order to survive. In this way, the interrelations between host and parasite often produce a multifaceted biological train of events, which is markedly more complex than the ecology of free-living species. In addition, many parasites develop in various habitats, for example, alternating between blood and tissue of a vertebrate host, and between the gut and salivary gland of an arthropod or a leech. The functional morphology is consequently linked with an often considerable change in the metabolism of the parasite. On this subject, many questions still remain unanswered. We are generally familiar only with the outward appearance of the course of an infection.

Parasites occur in all classes of the Protozoa. The Sporozoa are exclusively parasitic. It has been estimated that about one fifth of all Protozoa are parasites. If they live externally on their hosts, they are called **ectoparasites**. Ectoparasites occur on aquatic animals only because of the dependence of the Protozoa upon sufficient moisture in their environment. The endoparasites are distinguished as blood Protozoa or tissue Protozoa, or as parasites of the open body cavities, according to the organ system which they infect. The open body cavities include the entire digestive tract, the urinogenital system and the nasal cavities. Many parasites are capable of development in more than one focus of infection in their host and can infect several systems of organs, as can be seen from several of the following examples.

Parasitic flagellates

Among the flagellates of the Order Cryptomonadina, the genus *Chrysidella* contains species which live as symbionts in radiolarian Foraminifera, and also in sponges and actinae. They include *Chrysidella schaudinni* (figure 6 h) from *Peneroplis pertusus* (figure 40 b). The host provides a habitat for these zooxanthellae, and the parasite gives the host its assimilation products. The previously cited examples of the parasitic dinoflagellates *Haplozoon clymenellae* (figure 51–o) from the gut of polychaetes and *Blastodinium spinulosum* (figure 7 e–g), a parasite of copepods, show how greatly parasitism can modify not only the morphology but also the whole mode of development of Protozoa. Only the flagellate swarmers still retain the appearance of dinoflagellates.

The flagellates most important in medicine and economy belong to the Protomonadina. *Costia necatrix* (figure 17 a) is a skin parasite in fishes. The parasites, which adhere to the host by means of their flagellar depression, which acts as a sucking disc, can cause considerable losses in trout, carp and goldfish stocks, particularly in immature animals, by their massive multiplication, especially on the gills. The most important and the most interesting flagellates from a biological point of view are the Trypanosomatidae. The promastigote or leptomonad form living as a gut parasite in invertebrate hosts can be regarded as their basic type. *Leptomonas ctenocephali* from the gut of the dog flea (figure 17 b) is an example of this. The epimastigote or crithidial form (figure 17 c) has the kinetoplast close to the nucleus. Transmission occurs by the excretion of the parasites in the host faeces, during which amastigote stages also appear, an example of polymorphism (figure 23 b). If the arthropods concerned are blood suckers, the possibility arises that the parasite may also become adapted to the animal which has yielded the blood. Numerous examples show that this adaptation has been realized in a variety of forms. In the gut of the sheep ked *Melophagus ovinus*, epimastigote forms develop in large numbers. When the sheep ingests the infected ked from its fleece, the parasites reach the blood of the sheep via the mucous membranes, and are there transformed into the trypanosome form by displacement of the kinetoplast to the posterior end of the cell. These trypanomastigote stages of *Trypanosoma melo-*

phagium are, however, apparently able merely to survive in the blood of the vertebrate host, and are unable to reproduce there. In the morphologically similar but larger species *Trypanosoma theileri* (figure 17 d), which in the epimastigote form lives principally in the gut of horseflies (Tabanidae), the parasite multiplies in the blood of cattle especially at the onset of infection. Copious multiplication in the blood of the vertebrate host can be seen in the case of *Trypanosoma lewisi*, a parasite of rats which is transmitted when the rats ingest infected fleas. Only the development of an antibody inhibiting division, ablastin, stops multiplication in the blood of the rat.

Trypanosomes are a group of parasites widespread in the animal kingdom. They display a wealth of different species in the blood of fishes, amphibians, reptiles, birds and mammals, and generally cause non-pathogenic infections. The invertebrate hosts are mainly blood-sucking arthropods or leeches. Only the genus *Phytomonas* which morphologically resembles *Leptomonas* species, develops in the sap plants, mainly *Euphorbia* species, being transmitted by bugs.

The trypanosomes which can be classified in the *Trypanosoma brucei* complex are very important medically and economically. They are also of biological interest. *Trypanosoma brucei* lives in the trypomastigote form in the blood of African wild animals such as antelopes, zebras, buffaloes and elephants, without any detectable pathogenic effect. This adaptation between host and parasite does not occur in the domestic animals which have been introduced more recently. The infection is often lethal, in particular in horses, donkeys, camels, dogs and cats. Cattle and pigs, and more particularly sheep and goats, are more resistant to infection. The infection does not develop in man. Infection proceeds via flies of the genus *Glossina* (tsetse flies). After uptake of infected blood there is at first a multiplication of the parasites in the gut of *Glossina*, followed by their migration into the salivary glands; there at first they multiply further in the epimastigote stage, and finally change back into an infective trypomastigote form which will be transmitted to the vertebrate host when it is bitten by the insect. In contrast to *Trypanosoma melophagium*, *T. theileri* and *T. lewisi*, transmission by *Glossina* is by means of the insect bites, which is also true for the transmission of *Phytomonas* by bugs. Trypanosomes which are transmitted by insect bites together with the saliva are distinguished as **Salivaria** from the **Stercoraria** which are transmitted by the faeces of the vector. In the biological sense the *Stercoraria* are natural parasites of their invertebrate hosts, while the Salivaria represent a later adaptation effected by parasitism. Hence *Glossina* are not the primary hosts for *T. brucei*, which is why the infection develops in only a small percentage of *Glossina* individuals. *Glossina* flies support the natural mode of development, i.e. posteriorly through the gut, of *Trypanosoma grayi* which infects crocodiles. In this case the crocodiles are infected via the mucous membranes of the mouth by means of fly faeces or by ingestion of pieces of flies.

Apart from the economic losses in domestic animals, *T. brucei* infections merit interest since two species or subspecies which conform to *T. brucei* morphologically are also infective to man; these are *Trypanosoma gambiense* (figure 18 c) in west equatorial Africa and *T. rhodesiense* in east equatorial Africa. Both cause **African sleeping sickness** and are, like *T. brucei*, transmitted via the bite of *Glossina*. With them, too, only about 1–5% of the tsetse flies are infectious. If the natural alternation between *Glossina* and the vertebrate host is interrupted for a long time by repeated passage through the blood of the vertebrate host alone, the capacity to develop in the arthropods can be completely lost. This fact is reminiscent of the extrakaryotically determined state of long-lasting modification in streptomycin-requiring strains of *Chlamydomonas rheinhardi*, which depends on the genes of the mitochondrial DNA. In trypanosomes, the kinetoplast DNA is apparently important. Strains which have lost this DNA are unable to develop in the arthropod host. The phylogenetic transformation from *T. brucei* to *T. gambiense* and *T. rhodesiense*, respectively, probably occurred at different times. In *T. gambiense* there is already a better adaptation between man and the parasite. The course of the disease is less virulent than in *T. rhodesiense* infections. In addition *T. rhodesiense* infections have their reservoir in African wild animals, just as do the primitive *T. brucei* infections. A species of antelope, the bush buck, and some other wild animals are known to be a carrier for *T. rhodesiense* infections.

T. gambiense infection in man is demonstrable in the blood after about one week. There is periodic alternation of parasitic multiplication, with subsequent destruction by specific host antibodies, in which the parasite changes its antigenic character each time at the next multiplication phase in order again to evade the action of the antibodies. In this way a succession of fever periods is formed. Passage into the lymphatic system results in swelling, particularly of the cervical lymph nodes, as a typical symptom. Finally, the central nervous system is also infected. After periods of excitation, toxic damage induces somnolence, and the victim is soon no longer able to take food by himself, unless the course of the infection is interrupted by the nowadays very effective chemotherapy.

Adaptation to other vectors leads to trypanosomes of the *T. brucei* type occurring beyond the central African *Glossina* region. Trypanosomes which are transmitted by other flies, by Tabanidae, *Haematopota*, *Chrysops* and *Stomoxys*, are classified as *Trypanosoma evansi*. A rapid succession from animal to animal gives a purely mechanical transinoculation of blood during the act of biting. *T. evansi* is particularly pathogenic for horses, mules and camels, and occurs not only in central Africa but also in other parts of the world; in Madagascar, the Middle East, India, Thailand, Java, South America and Central America. The disease known as **surra** has a lethal course with involvement of the central nervous system. A further parasitological adaptation has occurred in *Trypanosoma equiperdum*. This species of trypanosome produces an infection of the genital mucous membranes in horses, donkeys and mules, in addition to dissemination in the vascular, lymphatic and central nervous systems. Thus these trypanosomes no longer require an arthropod vector. The disease becomes a venereal disease, **stallion cover disease** or **dourine**, which is spread by coitus. *Trypanosoma equinum* represents a further modification. In contrast to *T. evansi* and *T. brucei*, the central DNA constituent in the kinetoplasts is always absent in this species. The parasite is transmitted mechanically by the bite of Tabanidae and, in South America, also by the bite of blood-sucking vampire bats. The disease is called **mal de caderas** or **azoturia** and develops particularly in horses in the blood and central nervous system. Among wild animals, river swine also succumb to the infection. All the trypanosomiases discussed here reveal the extent of change which a parasite can accomplish phylogenetically.

Nagana, a disease of economic animals in Africa, is caused by *Trypanosoma vivax* and *Trypanosoma congolense*, which differ morphologically from *T. brucei*. These too are transmitted by *Glossina*. However, they have not become adapted to the salivary glands of the tsetse flies which carry them. *T. congolense* first infects the middle gut and then passes into the piercing proboscis; *T. vivax* infects only the proboscis.

Unlike the examples cited so far, in which the trypomastigote form of the parasite occurs always in the vertebrate host, *Leishmania* species assume the amastigote form in these hosts (figure 18 d), while multiplication in the gut of the insect vector occurs in the promastigote flagellate form (figure 17 b). *Leishmania* species likewise cause diseases in man. *Leishmania tropica*, which has spread to the Mediterranean region, is a skin parasite which causes a local abscess, **oriental sore**, at the site punctured by small *Phlebotomus* flies (Psychodidae, sandflies). Infection by *Leishmania braziliensis*, which occurs in South and Central America, follows a similar course, but is more protracted and occasionally shows a tendency to spread to the mucous membranes of the mouth and nose. *Leishmania donovani* (figure 18 d) which causes the severe disease **kala-azar**, mainly in India and the Mediterranean region, develops as a tissue parasite in the bone marrow, and also in the spleen and liver where it causes considerable enlargement. Transmission of the promastigote *Leishmania* by *Phlebotomus* proceeds mainly when abundant multiplication of the parasites in the stomach of the insect impedes the uptake of blood; the parasites are then vomited out.

Another route of infection of the vertebrate host is shown in the infection of man by *Trypanosoma cruzi* which causes **Chagas' disease** in South and Central America. *T. cruzi* is transmitted by large nocturnally sucking predatory bugs (Reduviidae) of the genera *Triatoma* and *Rhodnius*; infection of man is achieved by means of the faeces of the bugs via the wound left by the bite. The peculiarity in the development of this parasite in man consists firstly in the occurrence in the blood of the trypomastigote stage which is here, however, incapable of division, and secondly in the periodic transition to the amastigote form, mainly in the musculature, including cardiac musculature; it is in this latter stage that multiplication takes place.

While the Trypanosomatidae are parasites of the blood and tissues of the vertebrate host, the remaining parasitic flagellates cause infection of the open body cavities, mainly the gut. By far the most widely spread gut flagellates are the Trichomonadidae (figure 19), whose species abound as commensals in fishes, amphibians, reptiles, birds and mammals. They inhabit the regions of the gut which are colonized by bacteria, i.e. the colon, caecum and lower gut and, in ruminants, the rumen; they also infect the buccal cavity. As with all gut Protozoa, the Trichomonadidae develop more abundantly when their host eats food rich in plant matter than when it eats meat. Plentiful starch in the food encourages protozoan development. One host species can often be infected by several *Trichomonas* species. For example, 5 morphologically distinct species infecting man have been described; of these, one is an oral parasite and another, *Trichomonas vaginalis* (figure 19 d) infects the urinogenital tract. This infection by *T. vaginalis* can sometimes cause inflammation of the vagina and discharge in women. In men, the infection is usually latent. It is transmitted by the sex act. The urinogenital infection of cattle by *Tritrichomonas foetus* (figure 19 c) is markedly more pathogenic; a spread of the parasitic infection can lead to involvement of the internal organs of the foetus and hence to abortion. Since this disease, too, is transmitted by coitus, only *Tritrichomonas*-free bulls should be used so that an epidemic may be avoided.

A third *Trichomonas* species which can be pathogenically active is *Trichomonas gallinae* (*T. hepatica*), a parasite of the buccal cavity and crop mainly in poultry. Diphtheria-like .exudates occur in the crop region, particularly in young pigeons; the parasites spread into numerous internal organs in which necrotic lesions form during the lethal course of the disease. The Trichomonadidae thus provide a parallel to the Trypanosomatidae in which, too, only a few pathogens are found among numerous apathogenic species.

With regard to other parasitic flagellates belonging to the Protomonadina, reference should be made to figures 1 a, 17 g, h, 18 f, g and 19 a, b, e. *Giardia* species hold a special position as parasites of the small intestine. An example of this is *Giardia intestinalis* (figure 18 h), a parasite of man which can adhere to the villi of the small intestine by means of a fibrillar sucking disc (figure 117 a) and occasionally leads to symptoms of dysentery. The group of gut flagellates of the greatest biological interest are the numerous Polymastigina which occur in xylophagous termites and cockroaches (figure 20, 21); these live in close **symbiosis** with their host. The termites provide the parasites with a habitat, without which the parasites cannot develop. When the temperature or oxygen content are temporarily raised, the termites are rid of the flagellates but eventually also perish, since they lack cellulase to digest the food they consume. They need the Polymastigina with those cell substances and metabolic products they nourish themselves.

Parasitic Rhizopoda

Among the rhizopods, numerous amoebae and the Piroplasmidae live parasitically. Just like the Trichomonadidae, the parasitic amoebae are widespread commensals in the gut lumen colonized by bacteria. They occur in arthropods, leeches, fishes, amphibians, reptiles, birds and mammals, and a single host species can be infected by several species of amoebae. *Entamoeba histolytica* (figure 1 e, 29 e), *Entamoeba hartmanni, Entamoeba coli, Dientamoeba fragilis* (29 d), *Endolimax nana* (29 c) and *Iodamoeba bütschlii* (29 b) have been found in the gut of man and *Entamoeba gingivalis* in the human mouth. And, as in the Trichomonadidae, only a few species of amoebae are pathogenic. Of these, the principal is *Entamoeba histolytica* which causes **amoebic dysentery**. The infection of man by *E. histolytica* shows very clearly that even infection with a pathogenic parasite by no means always leads to illness. As in the other gut amoebae, the oral uptake of the cysts is followed by an infection of the gut lumen, with multiplication of the vegetative gut-lumen stage which is about 15 μm, from which cysts can again be produced (figure 156, lower half). The passage of the amoeba into the gut wall occurs when the man spends time in a warmer climate, if other additional factors, e.g. toxic action of bacteria, have reduced the resistance of the gut wall. Histolysis of the gut tissue then leads to the production of localized ulcers in the gut wall, which can be attributed to the secretion of proteolytic enzymes by the amoebae. Under these altered environmental conditions, the amoeba grows to 20–30 μm and becomes the tissue stage or magna form, which readily phagocytoses erythrocytes (figure 156 above, figure 1 e). The virulence of the amoeba hence evidently relates to the tendency to secrete proteolytic enzymes for extracellular digestion, as opposed to the intracellular digestion in the food vacuoles which otherwise occurs. The

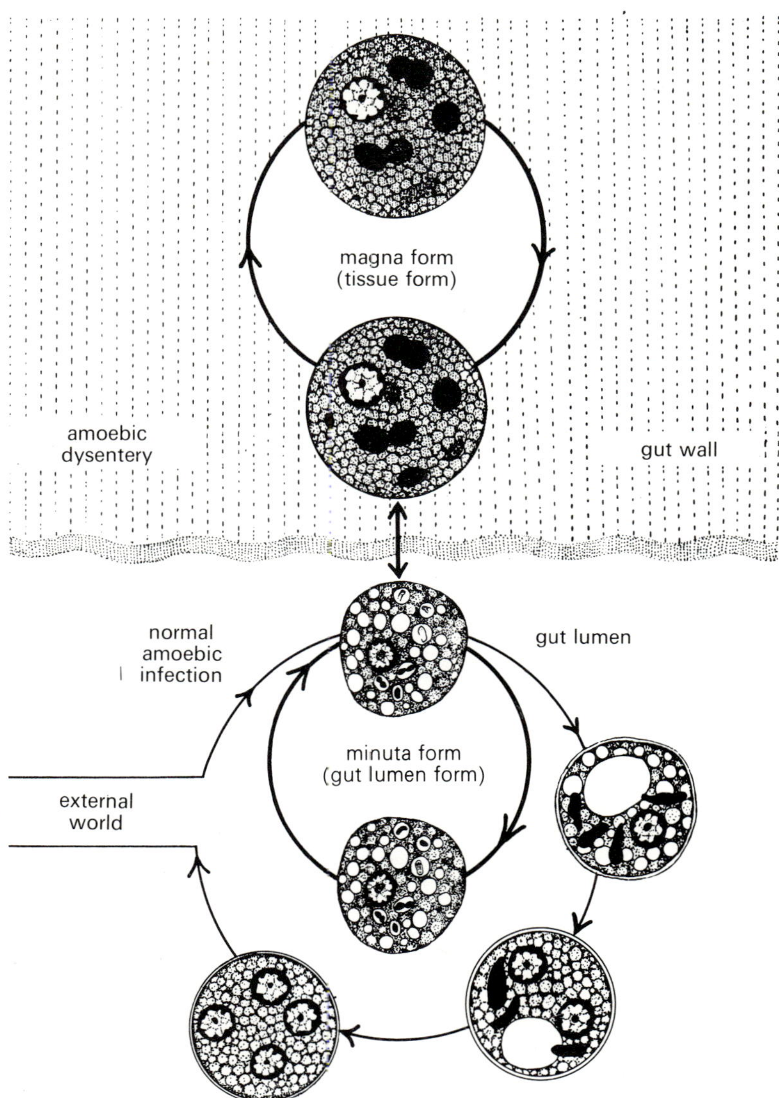

Figure 156. *Entamoeba histolytica*, **development in the gut lumen** (carrier of amoebae), and in the gut tissue (amoebic dysentery)

parasites can spread from the gut ulcers arising in amoebic dysentery via the vascular and lymphatic systems; this can lead to metastases, particularly in the liver, to new foci of infection which, with necrosis of the tissue, can develop into extensive abscesses. Like malaria, amoebic dysentery is a most important protozoan disease of man. It can sometimes be seen also in anthropoid apes in zoological gardens. An antithesis to the amoebic dysentery of man is the amoebic dysentery of reptiles caused by infection by *Entamoeba invadens*. Although it is generally a symptomless gut infection in turtles, *E. invadens* is highly pathogenic for snakes and lizards; in these animals the infection, which is nearly always lethal, involves the gut wall and as a rule the liver, but can also affect other organs such as the kidneys and pancreas. In these circumstances the turtles can be regarded as the true hosts of *E. invadens*.

Even normally free-living amoebae like the genera *Hartmanella* or *Acanthamoeba* (figure 29 a), and the mastigamoeba *Naegleria* (figure 17 k, l), can lead to infection and severe illness in man. The amoebae, taken up through the nose during bathing and swimming, can be seen primarily in the nasal mucus. The further course of the infection can involve the central nervous system. The inflammatory processes which set in in this region have the appearance of an acute meningoencephalitis. The infection, which then usually leads to death in a few days, can also be demonstrated in the cerebrospinal fluid during life.

All the Piroplasmidae are parasitic. They have often been assigned to the malaria parasites because they occur in the erythrocytes of the vertebrate hosts, but sexual processes and the ensuing sporogony are absent. *Babesia*, which are usually pyriform amoeboid parasites of the blood corpuscles, reproduce by binary fission (figure 30 a–c), and more rarely by dividing into four, after which the erythrocyte disintegrates. This leads to fever in the host animal, with high-grade anaemia and secretion of haemoglobin in the urine (haemoglobinuria). *Babesia bovis* causes **haemoglobinuria** of cattle mostly in Southern Europe, while in Northern Europe, including Britain, the similar but smaller type of *Babesia divergens* is the virus. **Texas fever** caused by *B. bigemina* (figure 30 a–c), which is found in warmer countries, is a more important disease. Other domestic animals and wild animals can also have *Babesia* infections. These are transmitted by ticks with primary development of the parasite in the gut wall. The subsequent infection of the eggs renders the next generation of ticks infectious, and transmission occurs when the tick bites. *Theileria* are primarily parasites of the lymphatic system in the vertebrate host. They reproduce also by schizogony. After infection of the lymphocytes (figure 30 d), parasitic stages finally occur in the erythrocytes (figure 30 e); after they have been taken up by a suitable tick, the parasites break through the gut wall into the salivary glands, where they multiply and are transmitted by the tick bite in the next tick stage. Economically the most important pyrexial disease, usually fatal, is the **African East Coast fever** in cattle, caused by *Theileria parva*. However, other *Theileria* species can also produce infection and illness in cattle or other animal species. Although man has a natural resistance towards the Piroplasmidae, *Babesia bovis* infections in man have nevertheless been known to lead to death in patients with spleen deficiency.

Sporozoa

All Sporozoa are parasites. The biological relationships connected with their development have already been discussed in detail in the taxonomic section, so that here it is chiefly their clinical effects which will be presented. Although the gregarines (figures 46–52) often occur in large numbers in their invertebrate hosts, no deleterious effects of the infections have been observed. The situation is different in the case of the intracellular Coccidia; severe infections lead to the destruction of numerous host cells, whereas milder infections do not cause any detectable illness. Some Eimeridea in particular can produce grave symptoms in the infected animals when external conditions encourage severe infections. Since these enter orally by means of the oocysts (figure 56), pools of standing water in the pasture can lead to the development of bloody diarrhoea in cattle caused by *Eimeria zürni* (figure 56 f) and the animals can die after 5–10 days. Other Coccidia of cattle (e.g. figure 56 g, h) are less pathogenic. The species *Eimeria tenella* can cause fatal dysentery in young birds in intensive poultry units. Infection of the bile duct in the liver of domestic rabbits by *Eimeria stiedae* (figure 56 a, b) often causes numerous losses, while the Coccidia of the gut (figure 56 c–e) are of less importance. Quantities of oocysts can sometimes gather in the air bladder of fishes, leading to an inability to swim, and hence to death. The Coccidia are parasites widely distributed in arthropods, fishes, amphibians, reptiles, birds and mammals. One host species can often be infected by several species, whereas all Coccidia species are very host-specific. In man, the species *Isospora belli* usually causes only mild disorders of the gut. In contrast to the Eimeridea, however, the Haemosporidia are of great pathogenic importance in man since they cause **malaria**.

The development of malaria parasites has already been shown in detail in figures 60 and 140. Figure 157 shows the connection between the clinical effects and the stage of development of the parasite during malarial infections in man. Schizogony in the blood decides the clinical pattern of fever. After the completion of schizogony, the infected erythrocytes disintegrate and release the residual body of the parasite, its metabolic products and the residual erythrocytic substance; this precipitates fever. In infection by *Plasmodium malariae* (**quartan malaria**, figures 157 a, 61 a–d) the largely synchronous timing of schizogony and the 72-hour duration of schizont development results in a bout of fever every third day. In *P. vivax* infections (**benign tertian malaria**, figures 157 b, 61 g–k) this occurs every 48 hours. The same is true for *P. falciparum* (**malignant tertian malaria**, figures 157 c, 61 e–f) except that here each bout of fever lasts longer because there is less synchronization. Malaria is the most important disease caused by Protozoa in man. As a function of the prevalence of the vectors, *Anopheles* mosquitoes, the disease has made large areas of swamp uninhabitable for man, and has decided the outcome of numerous wars. The geographical distribution of malaria is, however, determined not only by the presence of the vector mosquitoes but also by the temperature, which is critical for the development of plasmodia in the arthropod host. This development with the passage of the parasites into the salivary glands of the mosquitoes indicates that these cannot be considered to be primary hosts of plasmodia, since the natural development of the Coccidia in the arthropods, too, is directed posteriorly in the gut (figure 57).

The most dangerous type is falciparum malaria in which the schizonts can block the blood capillaries, particularly in the brain, resulting in coma, a deep unconsciousness. Effective drugs

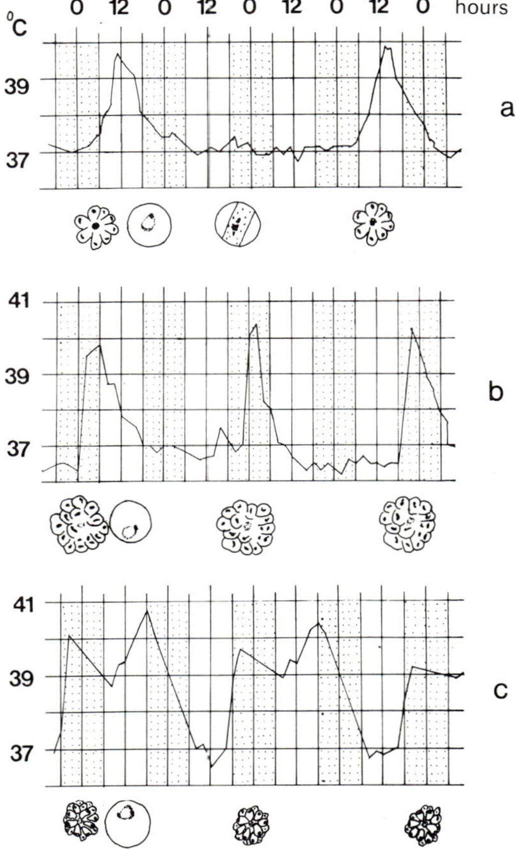

Figure 157. Dependence of the onset of fever in malaria upon the erythrocytic schizogony of each of the causal agents
  a *Plasmodium malariae*
  b *P. vivax*
  c *P. falciparum*

are available to combat the infection of the blood and to control the disease prophylactically; however, the primary infection in the parenchyma cells of the liver (figure 60 f, g) is more difficult to control and can therefore, especially in *P. vivax* and *P. malariae* infections, act for a certain time as the source of recrudescences in the blood.

In the same way that the pathogenic effects of the malaria species in man are diverse, so the many other haemosporidial infections in reptiles, birds and mammals differ widely from one another. In these the Haemosporidia, like the Coccida, have a marked host specificity, so that at best transmission is possible only to nearly related species of hosts. This is true also for the coccidial cycle of *Toxoplasma gondii* in cats of the genus *Felis* (figure 62 a–q). This unusual parasite shows a lack of organ- and host-specificity (figure 62 r–t) only with regard to the secondary host; here, numerous mammals, and also birds and reptiles, can be infected. In women, intrauterine transmission of the infection can result in abortion or in serious disease of the foetus. The affinity of the parasite for the central nervous system can cause hydrocephalus, encephalomyelitis and chorio-retinitis. In adults, too, **toxoplasmosis** infection via the mouth can lead to encephalomyelitis and also to pneumonia, hepatitis, etc., although by far the majority of these infections are latent. The infection of the musculature of many animals and occasionally also of man by Sarcosporidia (figure 64), which are closely related to *Toxoplasma*, is unimportant in comparison. Muscles of the pig which have been infiltrated extensively by *Sarcocystis miescheriana* show calcification after the death of the parasites, and the meat is condemned.

Among the Cnidosporidia, many Myxosporidia (figures 65–67) can produce a fatal infection in fishes. Should the host survive the extensive lesions in the musculature, e.g. in **boil disease** in barbels caused by *Myxobolus pfeifferi* (figures 66, 67), the fish are rendered unfit for consumption. Among the infections of arthropods by Microsporidia, **pébrine disease** of silkworms caused by *Nosema bombycis* (figure 70 a) and **bee dysentery** caused by *Nosema apis* are of particular economic importance. In this group, only *Pneumocystis carinii* (Haplosporidia, figure 72) infects man, mainly new-born babies. As has been described previously, this infection can cause the death of the infant by asphyxiation, with the symptoms of **interstitial pneumonia**.

Parasitic ciliates

The protociliates of the Anura (figure 73 a–d) are obviously commensals without clinical effects. Among the holotrichous ciliates, *Chilodonella cyprini* (similar to figure 74 a) can aggravate skin damage which has occurred in carps and goldfish, while swarmers of *Ichthyophthirius multifiliis* (figure 94 e–g) can penetrate even into the undamaged skin of fishes, thereby encouraging the development of bacteria and fungi, so that this type of infection can cause epidemics in fish stocks and aquaria. *Trichodina domerguei* (Peritricha, figure 87 c) causes epithelial proliferations on the skin, gills, and even in the urinary bladder, of trout and kills the fish brood. The Thigmotricha (figure 79) are parasites which fix their sucking proboscis deep ih the gills and the mantle cavity of their mussel host. However, *Spirochona gemmipara* (figure 90 a) uses the gills of *Gammarus pulex* solely as a substratum, and the host as a mode of conveyance; it is thus only an **epibiont**. The Astomata (figure 80) are mouthless gut parasites in oligochaetes, and apparently have no deleterious effects.

Nearly all ciliates which occur as gut parasites in terrestrial animals are harmless commensals; they occur chiefly in arthropods and vertebrates, e.g. the holotrichs *Paraisotricha colpoidea*, *Isotricha prostoma* and *Cyathodinium piriforme* (figure 76 b–d) or the spirotrich *Nyctotherus ovalis* (figure 82 h). The ciliate *Balantidium coli* (figure 1 c, d), also, is a harmless commensal in the caecum of its natural host the pig. *B. coli* can, however, occasionally be transmitted to man, who is therefore a secondary host for the species. In man, it again usually lives as a parasite of the gut lumen but, as with infection of man by the amoeba *Entamoeba histolytica*, certain factors can induce an involvement of the gut wall. *Balantidium* species cause destruction of the tissues, with the formation of ulcers, a disorder known as **Balantidium dysentery** which has led to death more quickly than does amoebic dysentery.

The amazing abundance of Entodiniomorpha (figure 83) in the rumen of ruminants leads us to wonder whether these specialized ciliates have the same significance as the Polymastigina in the xylophagous termites. However, the host thrives just as well after the removal of the ciliates with copper sulphate as it does when the ciliates are present. The breakdown of cellulose is bacterial in this case. The ciliates themselves are commensals, and depend upon the presence of available starch in the plant food, but at the same time serve as suppliers of protein for the host when the contents of the rumen are passed onwards.

Hyperparasitism and the parasites of the Protozoa

From several of the examples discussed, it can be seen that the parasitic Protozoa have sometimes not only become adapted to one host once and for all, but can also change to other hosts. Hyperparasitism provides evidence of this adaptation to new hosts; in this, the parasitic Protozoa become parasites of other parasites of the same host. *Entamoeba paulista*, similar to *E. ranarum* (figure 29f), is a parasite in the ciliate *Zelleriella braziliensis (Protoopalina)*. *Giardia muris* and *Trichomonas muris*, frequent gut parasites of the mouse, can also develop in the nematodes which parasitize the mouse. Various Microsporidia, too, develop in the trematodes and cestodes of their host. A species of *Hexamita* (Protomonadina, similar to *Octomitus*, figure 18g), a primitive gut flagellate, today flourishes more abundantly in a trematode of the eel than in its primary host. Thus a parasite of the gut lumen of the eel has become a parasite of the sexual organs of the worm, involving the uterus, oviduct, *receptaculum seminis* and the eggs.

That Protozoa can become parasitic in other Protozoa has already been shown in the example of the infection of *Zelleriella braziliensis* by *Entamoeba paulista*, as well as the development of *Tachyblaston ephelotensis* on *Ephelota gemmipara* (figure 95 e–o) and the infection of *Peneroplis pertusus* (Foraminifera, figure 40b) by *Chrysidella schaudinni* (Cryptomonadina, figure 6h). Various dinoflagellates can develop in Radiolaria and ciliates (*Tintinnidium*), just like the green zoochlorellae (algae) in amoebae, Testacea, Heliozoa and ciliates, e.g. in *Paramecium bursaria* (figure 77d). Most brownish zooxanthellae are encountered mainly in the marine Protozoa, in Radiolaria and Foraminifera. But fungi too (e.g. *Sphaerophrya*) and bacteria are parasites of the Protozoa; they include the kappa symbionts of killer Paramecium (figure 154) and the symbiotic bacteria of the amoeba *Pelomyxa palustris* (figure 27a) which were mentioned in connection with extrakaryotic inheritance and which have assumed the function of mitochondrial DNA. The question arises as to whether the DNA-containing cell organelles of the Protozoa—the chloroplasts, mitochondria and kinetoplast, together with the centriole—might have arisen originally as symbionts; however this still remains at present hypothetical. However, the fact that these organelles possess their own specific DNA is a very important argument for this supposition.

Bibliography

Books

Protozoology, including Parasitology and Techniques

Adam, H. and Czihak, G.: *Arbeitsmethoden*, Fischer, Stuttgart 1964.

Baker, J. R.: *Parasitic Protozoa*, Hutchinson University Library, London 1969.

Boch, J. and Supperer, R.: *Veterinärmedizinische Parasitologie*, Parey, Berlin–Hamburg 1971.

Brand, Th. von: *Parasitenphysiologie*, Fischer, Stuttgart 1972.

Bray, R. S.: *Studies on the Exo-erythrocytic Cycle in the Genus Plasmodium*, Lewis, London 1957.

Brown, H. W.: *Basic Clinical Parasitology*, 3rd ed., Butterworths, London, and Appleton-Century-Crofts, New York 1969.

Buetow, D. E.: *The Biology of Euglena*, Vol. 1/2, Academic Press, New York–London 1968.

Burck, H. C.: *Histologische Technik*, Thieme, Stuttgart 1966.

Doflein, F. and Reichenow, R.: *Lehrbuch der Protozoenkunde*, 6th ed., Fischer, Jena 1952/53.

Dogiel, V. A. (rev. Poljanskij, J. I, and Chejsin, E. M.): *General Protozoology*, Oxford University Press, London 1965.

Frank, W.: *Parasitologie*, Ulmer, Stuttgart 1976.

Garnham, P. C. C.: *Malaria Parasites and other Haemosporidia*. Blackwell Scientific Publications, Oxford 1966.

Garnham, P. C. C., Pierce, A. E.. and Roitt, I: *Immunity to Protozoa*, Blackwell Scientific Publications, Oxford 1963.

Göke, G. *Meeresprotozoen (Foraminiferen, Radiolarien, Tintinninen)*, Franckh (Kosmos), Stuttgart 1963.

Grassé, P. P.: *Protozoaires. Traité de Zoologie*, Vol. 1, Masson et Cie., Paris 1952/53.

Grell, K. G.: *Protozoology*, Springer, Berlin-Heidelberg-New York 1973.

Grospietsch, Th.: *Wechseltierchen (Rhizopoden)*, 2nd ed., Franckh (Kosmos), Stuttgart 1965.

Hall, R. P.: *Protozoology*, Prentice-Hall Inc., New York 1953.

Hammond, D. M. and P. L. Long: *The Coccidia*, University Park Press, Baltimore 1973.

Harnisch, O.: *Rhizopoda*, in Brohmer, P., Ehrmann, P., and Ulmer, G.: *Tierwelt Mitteleuropas*, Vol. 1, Part 1b, Quelle & Meyer, Leipzig.

Hartmann, M.: *Praktikum der Protozoologie*, 5th ed., Fischer, Jena 1928.

Hayat, M. A.: *Principles and Techniques of Electron Microscopy*, Vol. 1/2, Van Nostrand Reinhold Company, New York 1970/72.

Hess, D.: *Pflanzenphysiologie*, 2nd ed., Ulmer, Stuttgart 1972.

Hill, D. L.: *The Biochemistry and Physiology of Tetrahymena*, Academic Press, New York-London 1972.

Hoare, C. A.: *The Trypanosomes of Mammals*, Blackwell Scientific Publications, Oxford-Edinburgh 1972.

Huber-Pestalozzi, G.: *Das Phytoplankton des Süsswassers*, in Thienemann, A.: *Die Binnengewässer*, Vol. 16, Part 2/1, 3, 4, Schweizerbart (E. Nägele), Stuttgart 1941/50/55.

Jeon, K. W.: *The Biology of Amoeba*, Academic Press, New York-London 1973.

Jírovec, O.: *Parasitologie für Ärzte*, Fischer, Jena 1960.

Jones, A. R.: *The Ciliates*, Hutchinson, London 1974.

Koehler, J. K.: *Advanced Techniques in Biological Electron Microscopy*, Springer, Berlin-Heidelberg-New York 1973.

Kudo, R. R.: *Protozoology*, 6th ed., Thomas, Springfield (Illinois) 1971.

Liebmann, H.: *Handbuch der Frischwasser- und Abwasserbiologie*, 2nd ed., Vol. 1, Oldenbourg, München 1962.

Mackinnon, D. L. and Hawes, R. S. J.: *An Introduction to the Study of Protozoa*, Oxford University Press, London 1961.

Matthes, D. and Wenzel, F. *Wimpertiere (Ciliaten)*, Franckh (Kosmos), Stuttgart 1966.

Mayer, M.: *Kultur und Präparation der Protozoen*, 3rd ed., Franckh (Kosmos), Stuttgart 1956.

Pellérdy, L. P.: *Coccidia and Coccidiosis*, 2. Rev. Parey, Berlin-Hamburg 1974.

Piekarski, G.: *Lehrbuch der Parasitologie*, Springer, Berlin-Göttingen-Heidelberg 1954.

Piekarski, G.: *Medizinische Parasitologie*, 2nd ed., Springer, Berlin-Heidelberg-New York 1973.

Reimer, L.: *Elektronenmikroskopische Untersuchungs- und Präparationsmethoden*, 2nd ed., Springer, Berlin-Göttingen-Heidelberg 1967.

Richardson, U. F. and Kendall, S. B.: *Veterinary Protozoology*, Oliver and Boyd, Edinburgh-London 1957.

Romeis, B.: *Taschenbuch der Mikroskopischen Technik*, 16th ed., Oldenbourg, München-Wien 1968.

Ruthmann, A.: *Methoden der Zellforschung*, Franckh (Kosmos), Stuttgart 1966.

Sleigh, M. A.: *The Biology of Protozoa*, Arnold, London 1973.

Sleigh, M. A.: *Cilia and Flagella*, Academic Press, London-New York 1974.

Steinecke, F.: *Das Plankton des Süßwassers*, 2. Aufl., Quelle & Meyer, Heidelberg 1974.

Streble, H. and Krauter, D.: *Das Leben im Wassertropfen. Mikroflora und Mikrofauna des Süsswassers*, Franckh (Kosmos), Stuttgart 1973.

Trussel, R. E.: *Trichomonas vaginalis and Trichomoniasis*, Thomas, Springfield/Illinois 1947.

Van Wagtendonk, W. J.: *Paramecium, a Current Survey*, Elsevier, Amsterdam 1974.

Wenyon, C. M.: *Protozoology*, Vol. 1/2, Baillière, Tindall and Cox, London 1926.

Westphal, A.: *Protozen*, in Reichenow/Vogel/Weyer: *Leitfaden zur Untersuchung der tierischen Parasiten des Menschen und der Haustiere*, 4th ed., Barth, Leipzig 1969.

Westphal, A.: *Protozoa*, in Brohmer: *Fauna von Deutschland*, 11th ed., Quelle & Meyer, Heidelberg 1971.

Westphal, A. (with Mühlpfordt, H.): *Protozoen*, Ulmer, Stuttgart 1974.

Wetzel, A.: *Protozoa*, in Kaestner: *Lehrbuch der Speziellen Zoologie*, 3rd ed., Fischer, Stuttgart 1968.

Wichterman, R.: *The Biology of Paramecium*, Blakiston, New York-Toronto 1953.

Periodicals

Protozoology, including Parasitology

Acta Protozoologica, Warszawska Drukarnia Naukowa, Warszawa.
Acta Tropica, Schwabe, Basel.
Archiv für Protistenkunde, Fischer, Jena.
Mikrokosmos, Franckh (Kosmos), Stuttgart.
Protistologica, Centre National de la Recherche Scientifique, Paris.
Society of Protozoologists. A Revised Classification of the Phylum Protozoa.
  In *The Journal of Protozoology* 11, 7–20, 1964.
The Journal of Protozoology, Allen Press, ,Lawrence.
Zeitschrift für Parasitenkunde, Springer, Berlin-Heidelberg-New York.
Zeitschrift für Tropenmedizin und Parasitologie, Thieme, Stuttgart.

Sources of illustrations

All the illustrations have been redrawn to give maximum uniformity. The following alphabetical list shows which authors supplied the originals. Some drawings have been modified, simplified or brought into line with present-day knowledge. Thus all cycles of the Foraminifera and Sporozoa have been arranged in a uniform system. In the taxonomic section, individual figures generally contain illustrations at the same degree of enlargement, so that the relative sizes of the specimens may readily be judged. Where the original occurred in a work by more than one author, or where another author has made any material changes, the figure number appears against the names of all authors concerned.

Adie 59. Aikawa 140 2E–10E. Alexeieff 17 e, h, k, l, 64 f. Allegre 8 h. Allen 118 b. Antipa 128. Aragão 59.

Bathia 73 d. Becker 76 c, 83 a, e, f, 140 16E. Belar 44 f–i. Berndt 49 a–d. Biecheler 11 f. Bird 140 16E. Blochmann 31. 74 a, 82 k. Bold 12 e. Bolochonzew 2 k. Bonner 28 g. Borgert 2 o, 43 e, f, 104 d–f. Brady 39 c, 40 a, 41 a, c, 43 n. Brandt 43 g, 111 a. Braunitzer 143 a, b. Breslau 52 a. Bütschli 77 b, 84 e, 95 a–c, p–r, 123 g. Bullington 123 b–e. Bundle 83 h. Burck 16 a. Bushkiel 52 a.

Cachon 7 d. Cachon-Enjumet 7 d. Calkins 84 d, 94 a. Le Calvez 41 e, 43 p, q. Carpenter 40 b, c. Carter 4 a, c–e. Cash 36 a, 37 f, i. Cépède 80 a, 96 a. Chatton 7 e–g, 15 b, 25 a–i, 79 d, f, 81 a–c. Cheissin 79 c, 80 b, c. Chen 73 a, 92 a. Chodat 14 c. Christensen 56 f–l. Clemmons 129. Clermond-Ferrand 137 e. Cleveland 24 m–o, 142, 146 f–i. Collin 91 b, g, h, 95 s–u, 100 k–m. Conklin 80 d. O'Connor 18 f, 28 b, 29 a. Conrad 3 b. Craggs 118 a. da Cunha 9 a.

Dangeard 12 c. Deflandre 2 p, 9 k. Delain 113. Dellinger 124 d. Dill 13 a, 23 c, d, l–n. Diwald 25 k–z. Dobell 23 f, 24 a–c, 28 b, 29 a, f, 56 n, q–s, 58. Doflein 2 a, i, q, 3 c–f, 6 a, b, e, 9 b, 8 c, l, n, o, 16 g, 17 f, 25 j, 26 a, 27 a, 28 d, 32 b, 37 h, 43 a–c, h, i, 65 b, 94 b, c, h–k, 116 h–k, 124 a, 127 a, 135 e. Dogiel 83 b, d, g, 139 l. Doyle 124 f, Dragesco 75 f. Dubey 62 a–e. Dujardin 8 g, 42 a. Dunachie 62 f–g.

Van Eecke 64 e. Ehrenberg 135 a, b. Elliot 129. Enriques 99. Entz 17 a, 84 a. Erdmann 144.

O'Farrell 17 c, 23 h. Fauré-Fremiet 75 a, 84 b. Finlay 88 a. Förster 143 a, b. Foettinger 96 b–d. Fouquett 94 f, g. Frenkel 62 a–e. de Fromentel 89 c.

Index

Numbers in bold type indicate pages carrying the explanation of the respective technical terms. Page numbers with * refer to figures.

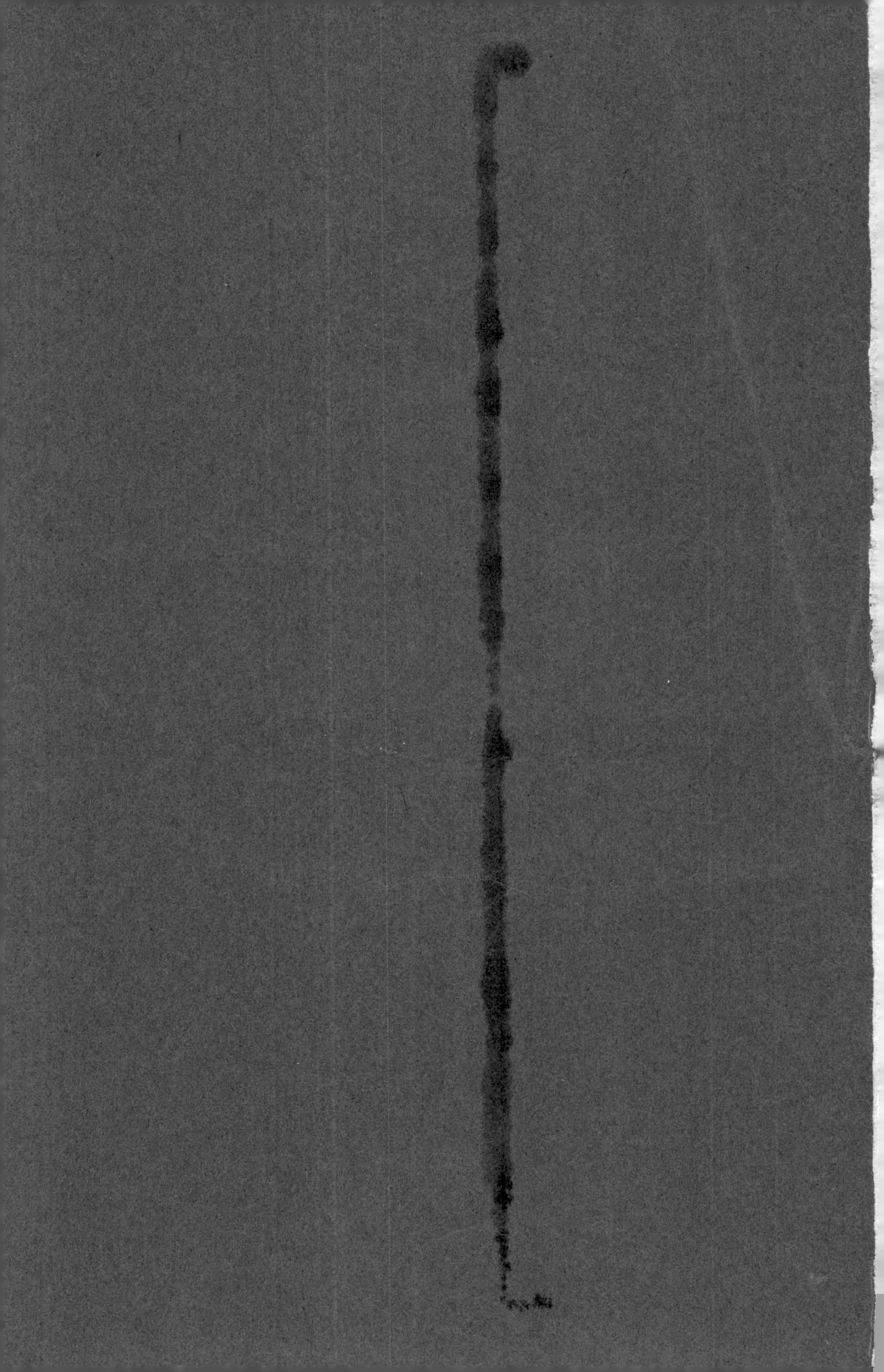